More Rapid Math Tricks and Tips

Other math titles by Edward H. Julius

Rapid Math Tricks and Tips: 30 Days to Number Power
Rapid Math in 10 Days: The Quick-and-Easy Program for Mastering Numbers
Arithmetricks: 50 Easy Ways to Add, Subtract, Multiply, and Divide Without a Calculator

More Rapid Math Tricks and Tips

30 Days to Number Mastery

EDWARD H. JULIUS

John Wiley & Sons, Inc.
New York • Chichester • Brisbane • Toronto • Singapore

This text is printed on acid-free paper.

Published by John Wiley & Sons, Inc.

Library of Congress Cataloging-in-Publication Data:
Julius, Edward H.
More rapid math tricks and tips: 30 days to number mastery/
Edward H. Julius.
p. cm.
ISBN 0-471-12238-6 (paper : alk. paper)
1. Arithmetic. I. Title.
QA111. J84 1996
513.9 — dc20 95-35581

Printed in the United States of America

10 9 8 7 6 5 4 3 2 1

To my parents, Nathan and Eleanor Julius

Acknowledgments

I would like to gratefully acknowledge my editor, Judith McCarthy, for her expert guidance and for once again believing in me. In addition, my hat goes off to John Cook for ably overseeing the book's production. Next, special thanks are due to Professor Jack Chapman for serving as an important source of information. Finally, I wish to acknowledge my friend and mentor, Dr. Lyle Sladek, whose input and encouragement have meant more to me than words can express.

— Contents —

For your enlightenment and enjoyment, a number of mathematical curiosities, tips, anecdotes, and challenges have been interspersed throughout this book.

— Introduction —

What would life be without arithmetic?
—Sydney Smith (1771–1845)

First and foremost, without arithmetic there would not be this fascinating and useful book!

All kidding aside, let me guess what you wondered as you picked up this book for the first time and read its title: *More Rapid Math Tricks and Tips* — You wondered if it is necessary to have read the original *Rapid Math Tricks and Tips* book to be able to understand this book, the sequel. The answer is an emphatic "No!" I would never be foolish enough to write a book that requires prerequisite reading. There are actually only two tricks from the original book that you need to know for this one (and for just a handful of the latter tricks), and these are reproduced in an abridged form in Appendix A of this book.

On the other hand, if you want to become a mental math monster by learning all the good tricks in existence, you may wish to obtain a copy of the original *Rapid Math Tricks and Tips* (which can be purchased or ordered at finer bookstores everywhere). A list of the 60 tricks presented in that book comprises Appendix B of this book.

Well, now that I got my book plug out of the way, let me talk about how you will benefit from this book. In short, you will learn how to work addition, subtraction, multiplication, and division problems faster than you thought possible, in most cases in your head.

Hold on There. I Just Let My Calculator Do All the Work.

Don't deny it — you just thought that, didn't you? Well, I guess it's time for me to convince you that you should not be a slave to your calculator and should be able to perform most calculations in your head. Here are my arguments:

- Calculators do break down from time to time. When this happens to my students during an exam, they stare at me in disbelief when I suggest that they use pencil and paper!
- Think of how embarrassed you'd feel (or should feel) if you were asked to multiply 5 by 12 and had to pull out your calculator to get the answer. Now think of how proud you'd feel if you were asked to divide 128 by 16 and came up with the answer almost instantly!
- Sometimes you'd prefer to perform a calculation secretly. Let's say you think a store clerk has overcharged or shortchanged you. It will embarrass the clerk, and consume too much time, if you decide to redo the math on your calculator.

- Sometimes it's awkward or inconvenient to use a calculator, as when you're driving (now that would be dangerous) or your hands are full. Many calculators that are convenient enough to carry around are so small that you could go blind trying to use them!
- Sometimes you're simply not allowed to use a calculator, as when taking certain standardized exams.
- When engaged in rapid-fire negotiations, at work or even when buying a car, you will definitely gain the upper hand if you can perform all the math in your head.
- Many calculations are actually faster to do in your head than on a calculator — once you know the tricks!

So, the next time it's your turn to calculate the tip at a restaurant, or you only have $25 on you and an unknown cost of groceries in your shopping cart, you won't have to break into a cold sweat. You'll have the confidence and ability to overcome the situation by performing the tricks you are about to learn. In short, you'll learn to love working with numbers.

But I'm Not a Direct Descendant of Albert Einstein.

You don't have to be. All you need is a basic understanding of addition, subtraction, multiplication, and division. That's all! For those of you who are a bit rusty at math, I've provided a review of basic math concepts and terminology in the pages that follow. What is the primary prerequisite for this "number-mastery" program? Simply the desire to learn rapid math. You obviously have that desire, because you were curious enough to buy this book!

I'm Intrigued. Tell Me Some More about Rapid Math.

Okay, if you insist. As you work through this number-mastery program, please keep in mind the following:

- You must practice, practice, practice these tricks to master them. Rapid math cannot be learned overnight (then again, what of value can?). On the other hand, you'll have plenty of fun along the way!
- You probably have already used some math tricks without realizing it. I'll never forget the time my daughter was just learning addition, and I saw her add $8 + 7$ in almost no time. When I asked her how she got the answer so quickly, she said, "Well, I know that $8 + 8 = 16$, so $8 + 7$ must equal one less than 16." She had just used a mental math trick.
- The more tricks you master, the more often you'll be able to perform mental math. But don't expect to learn all the tricks in one day or even in

one week. If you only learn the two per day (in most cases) contained within this 30-day program, that will be plenty. On the other hand, you need not master all the tricks—just use the ones you like and can remember the easiest.

- A trick is beneficial only if it is not too cumbersome and is truly faster than the usual method. It must also be practical. So the 13-step trick to calculate division by 736 did not make it into this book.
- Because addition and multiplication are, for some strange reason, generally faster and easier to execute than subtraction and division, you will be guided in that direction whenever possible.
- Don't hesitate to use more than one trick simultaneously or perhaps a variation I have not addressed, if you think it will help. In the end, your primary goal should be to use whatever technique(s) will produce the correct answer in the quickest and easiest way.
- You'll need to hide your calculator immediately! I'll pause for a moment while you do that.

Okay, I've Disowned My Calculator. What Do I Do Now?

Good, it sounds like you're ready to get down to business. Here are my recommendations for getting the most out of this number-mastery program.

First, finish reading the introduction, including the review section. You'll then note that the book is comprised of 54 rapid math tricks, plus four dynamite "parlor tricks."

The 54 tricks are carefully organized into basic, intermediate, advanced, and unusual tricks (as well as a trick to check your math on Days 29 and 30). Within each trick, you'll first read an explanation of strategy. Then you'll see two "Elementary Examples" and one "Brain Builder" example explained in step-by-step fashion. The Brain Builder examples basically involve larger numbers and more sophisticated applications.

Most examples are followed by a "Thought Process Summary" section, which provides a good conceptual overview of the technique. Additional insight into a trick is then provided in a "Food for Thought" section.

When you feel as though you understand the trick at hand, you're then ready to test yourself by working the practice problems. If all you want is a basic understanding of the trick, just work the "Elementary Exercises" and skip the "Brain Builders."

After many of the tricks, you'll see a "reward" section, consisting of some additional math tips, curiosities, and challenges. These extras have been provided for your enlightenment and enjoyment.

A "Quick Quiz" is provided at the end of each week, and a final exam is provided upon completion of the entire program. All solutions to practice problems and quizzes are provided at the back of the book.

Finally, Appendix A contains the two tricks explained in the original *Rapid Math Tricks and Tips* that you'll need to know for this book, and Appendix B lists the 60 tricks contained in that book. So if you are already aware of some good math tricks but cannot find them in this book, chances are they were already explained in the original volume.

Great, Ed! Now Can You Please Conclude the Introduction So I Can Get on with This Program?

Fair enough. Just fasten your seat belt, put on your thinking cap, and get ready to become educated, entertained, and empowered!

Ed Julius

Brief Review of Some Important Math Concepts

Place Value

This may seem basic, but it is important that you understand what each digit within a number stands for. For example, the numbers 437 and 374 contain the same three digits, but they do not represent the same amount because the digits have different *place values*.

In the number 437, the 7 is the ones digit. It equals 7×1, or 7. The 3 is the tens digit. It equals 3×10, or 30. The 4 is the hundreds digit. It equals 4×100, or 400. So the number 437 can be thought of as $400 + 30 + 7$. On the other hand, the number 374 equals $300 + 70 + 4$.

Place value also applies to decimals. Let's look at the number 5.29. The 5 is the ones digit and equals 5×1, or 5. The 2 is the tenths digit. It equals $2 \times \frac{1}{10}$, or $\frac{2}{10}$. The 9 is the hundredths digit. It equals $9 \times \frac{1}{100}$, or $\frac{9}{100}$. So the number 5.29 can be thought of as $5 + \frac{2}{10} + \frac{9}{100}$. It can also be described as $5 + \frac{29}{100}$.

Different Ways to Express the Same Number

There are certain ways a number can be written without changing its value. Let's take the number 58. It can be written as 58.0, 58.00, and so on and still mean 58. It can also be written as 058, 0058, and so on and still mean 58. (It is unusual, however, for a number such as 58 to be written as 058 or 0058 — there's no real need to put extra zeros in.)

On the other hand, putting the zero in-between or after the 5 and 8, giving you 508 or 580, would not mean the same as the number 58.

Numbers Can Be Added in Any Order

You can add numbers in any order to get the same sum. For example, the addition problem $32 + 19 + 66$ could be written as $66 + 32 + 19$ (or in four other ways) to produce the answer, 117.

Numbers Can Be Multiplied in Any Order

You can also multiply numbers in any order to get the same product. For example, the multiplication problem 8×13 could be written as 13×8 to produce the answer, 104.

What It Means to Square a Number

When you square a number, you are multiplying the number by itself. For example, 8^2 (spoken as "8-squared") is the same as 8×8, or 64.

Subtraction Is the Inverse of Addition

Subtraction can be thought of as the opposite of addition. For example, $14 + 7 = 21$. Turning things around, you could then say that $21 - 7 = 14$ and $21 - 14 = 7$.

Division Is the Inverse of Multiplication

Division can be thought of as the opposite of multiplication. For example, $6 \times 9 = 54$. Turning things around, you could then say that $54 \div 9 = 6$ and $54 \div 6 = 9$.

Division Can Be Shown in Three Different Ways

You can show a division problem in three different ways. For example, 13 divided by 2 can be written as $13 \div 2$, as $2\overline{)13}$, or as $\frac{13}{2}$. In school, you probably learned that $13 \div 2 = 6$ r1 (6, remainder 1). However, that answer is more commonly shown as 6.5 or $6\frac{1}{2}$.

How Decimals and Fractions Relate

A number might be a whole number, such as 7, 18, or 206, or it might fall between two whole numbers, such as 81.7 or $36\frac{3}{4}$. As shown, both decimals and fractions are used to show part of a whole. It's important to know what they mean and how a fraction can be expressed as a decimal (or the other way around).

Page 8 shows some common fractions and equivalent decimals. For example, the fraction $\frac{3}{4}$ is equivalent to 0.75. So a number such as $18\frac{3}{4}$ could also be written as 18.75.

Multiplying by 10, 100, Etc., Is Easy!

To multiply a number by 10, 100, and so forth, simply tack on the appropriate number of zeros to the right of the number, or move the decimal point the appropriate number of places to the right. So $34 \times 10 = 340$, and $87 \times 100 = 8{,}700$. Similarly, $6.02 \times 10 = 60.2$, and $9.25 \times 100 = 925$.

Dividing by 10, 100, Etc., Is Also Easy!

To divide a number by 10, 100, and so forth, simply move the decimal point of the number the appropriate number of decimal places to the left. So $28 \div 10 = 2.8$, and $733.4 \div 100 = 7.334$. Also, when dividing, you may cancel an equal number of right-hand zeros. So $900 \div 30 = 90 \div 3$, and $6{,}000 \div 300 = 60 \div 3$.

Symbols, Terms, and Tables

Symbols

1. "+" means "plus" or "add"
2. "−" means "minus" or "subtract"
3. "×" means "times" or "multiplied by"
4. "÷" means "divided by"
5. "=" means "equals"
6. "n^2" (spoken "*n*-squared") means a certain quantity multiplied by itself

Operational Terms

39	←	Addend
+ 22	←	Addend
61	←	Sum

Multiplicand	→	17	←	Factor
Multiplier	→	× 56	←	Factor
Product	→	952		

483	←	Minuend
− 291	←	Subtrahend
192	←	Difference

891	÷	33	=	27
↑		↑		↑
Dividend		Divisor		Quotient

Addition/Subtraction Table

+	**1**	**2**	**3**	**4**	**5**	**6**	**7**	**8**	**9**	**10**	**11**	**12**
1	2	3	4	5	6	7	8	9	10	11	12	13
2	3	4	5	6	7	8	9	10	11	12	13	14
3	4	5	6	7	8	9	10	11	12	13	14	15
4	5	6	7	8	9	10	11	12	13	14	15	16
5	6	7	8	9	10	11	12	13	14	15	16	17
6	7	8	9	10	11	12	13	14	15	16	17	18
7	8	9	10	11	12	13	14	15	16	17	18	19
8	9	10	11	12	13	14	15	16	17	18	19	20
9	10	11	12	13	14	15	16	17	18	19	20	21
10	11	12	13	14	15	16	17	18	19	20	21	22
11	12	13	14	15	16	17	18	19	20	21	22	23
12	13	14	15	16	17	18	19	20	21	22	23	24

Multiplication/Division Table

×	1	2	3	4	5	6	7	8	9	10	11	12
1	1	2	3	4	5	6	7	8	9	10	11	12
2	2	4	6	8	10	12	14	16	18	20	22	24
3	3	6	9	12	15	18	21	24	27	30	33	36
4	4	8	12	16	20	24	28	32	36	40	44	48
5	5	10	15	20	25	30	35	40	45	50	55	60
6	6	12	18	24	30	36	42	48	54	60	66	72
7	7	14	21	28	35	42	49	56	63	70	77	84
8	8	16	24	32	40	48	56	64	72	80	88	96
9	9	18	27	36	45	54	63	72	81	90	99	108
10	10	20	30	40	50	60	70	80	90	100	110	120
11	11	22	33	44	55	66	77	88	99	110	121	132
12	12	24	36	48	60	72	84	96	108	120	132	144

Some Squares You Should Know

$1^2 = 1$	$6^2 = 36$	$11^2 = 121$	$16^2 = 256$
$2^2 = 4$	$7^2 = 49$	$12^2 = 144$	$17^2 = 289$
$3^2 = 9$	$8^2 = 64$	$13^2 = 169$	$18^2 = 324$
$4^2 = 16$	$9^2 = 81$	$14^2 = 196$	$19^2 = 361$
$5^2 = 25$	$10^2 = 100$	$15^2 = 225$	$20^2 = 400$

Table of Equivalencies

Fraction		Decimal Equivalent	Fraction		Decimal Equivalent
$\frac{1}{100}$	=	0.01	$\frac{3}{8}$	=	0.375
$\frac{1}{50}$	=	0.02	$\frac{2}{5}$	=	0.4
$\frac{1}{40}$	=	0.025	$\frac{1}{2}$	=	0.5
$\frac{1}{25}$	=	0.04	$\frac{3}{5}$	=	0.6
$\frac{1}{20}$	=	0.05	$\frac{5}{8}$	=	0.625
$\frac{1}{10}$	=	0.1	$\frac{2}{3}$	=	0.666 . . .
$\frac{1}{8}$	=	0.125	$\frac{3}{4}$	=	0.75
$\frac{1}{5}$	=	0.2	$\frac{4}{5}$	=	0.8
$\frac{1}{4}$	=	0.25	$\frac{7}{8}$	=	0.875
$\frac{1}{3}$	=	0.333 . . .	$\frac{n}{n}$	=	1.0

One Final Note Before You Get Started: Applying a Test of Reasonableness

Whenever you complete a calculation, you should always give your answer a quick check (or "test of reasonableness") to see if it seems to be in the ballpark. In other words, always check to see if your answer looks right. Let's say you've just added 93 and 98 and arrive at the answer 201. You know that each of the numbers being added is *less* than 100, so the sum of the two should be less than 200. Therefore, 201 is too high to be the answer (the sum is actually 191).

A test of reasonableness should also be done for multiplication, division, subtraction, and squaring. Sometimes it's difficult to apply this test, especially if the numbers involved are large. With practice, though, you'll be able to quickly spot an incorrect answer. When this happens, simply redo the calculation until it looks right.

Week 1 More Basic Rapid Math Tricks

Trick 1: Multiplying with Little or No Carrying

Strategy: We begin our number-mastery program with a trick that'll be right up your alley if you hate to carry (and who doesn't). It works best when multiplying by a one-digit number and involves the following procedure. First, multiply by the ones digit, and write down the product in the answer space. Then, multiply by the tens digit, but place the product underneath the previous product, one column to the left. Continue in this manner until you run out of digits to multiply. Finally, add the partial products, and you've got the answer! As you'll see below, it's far simpler than it sounds.

Elementary Example 1

38 × 7

```
                    38
                  ×  7
Step 1.  8 × 7 =    56
Step 2.  3 × 7 =   21
Step 3.  Add  →    266 (answer)
```

Elementary Example 2

64 × 9

```
                    64
                  ×  9
Step 1.  4 × 9 =    36
Step 2.  6 × 9 =   54
Step 3.  Add  →    576 (answer)
```

Brain Builder

716 × 6

```
                        716
                      ×   6
Step 1.  6 × 6 =         36
Step 2.  1 × 6 =        06  ← (Write "06" so placement of "42" is easier.)
Step 3.  7 × 6 =      4 2
Step 4.  Add   →    4,296 (answer)
```

Food for Thought: Adding, too, can be achieved without carrying, in a manner similar to the technique shown above. An interesting variation, however, will constitute our next trick, as we add with little or no carrying. Not only does carrying take longer than our technique above, it "carries" with it a much higher likelihood of error. Furthermore, it is easier to check your answer when you have followed a no-carry approach.

Practice Problems

It's your turn, but no "carrying-on" is allowed!

Elementary Exercises

```
1.   46        5.   98        9.   56       13.   53
   × 8            × 6            × 4             × 9
    48
   32
   368

2.   93        6.   84       10.   28       14.   39
   × 5            × 7            × 7             × 5

3.   59        7.   65       11.   43
   × 4            × 6            × 8

4.   77        8.   83       12.   79
   × 9            × 9            × 6
```

Brain Builders

1. 426 × 7	3. 374 × 8	5. 645 × 6	7. 269 × 7
2. 918 × 5	4. 997 × 9	6. 889 × 4	8. 741 × 8

(See solutions on page 199.)

Trick 2: Adding with Little or No Carrying

Strategy: Adding, like multiplying, can be achieved without carrying. We're going to present, however, a slightly different approach to entering the partial sums from the method shown in Trick 1. As you will see below, we're going to "bunch up" the partial sums so they will be easier to use in producing the total.

Elementary Example 1

375 + 814 + 266 + 903

Step 1. Add the ones digits: 5 + 4 + 6 + 3 = 18.

Step 2. Add the tens digits: 7 + 1 + 6 + 0 = 14.

Step 3. Add the hundreds digits: 3 + 8 + 2 + 9 = 22.

Step 4. Add the partial sums, as shown below.

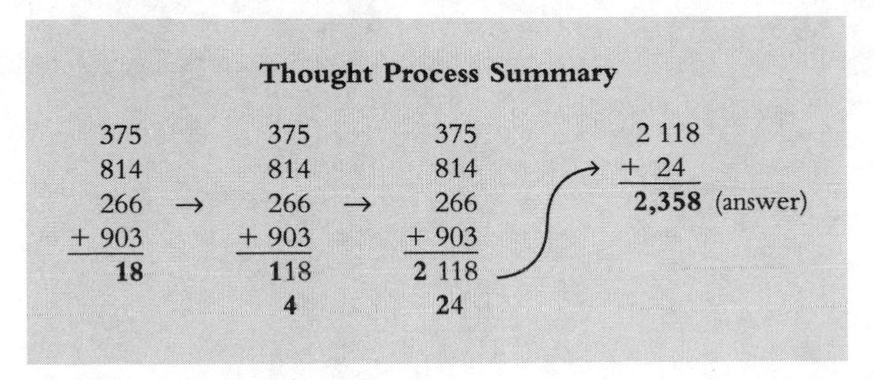

Elementary Example 2

791 + 246 + 385 + 504

Step 1. Add the ones digits: $1 + 6 + 5 + 4 = 16$.

Step 2. Add the tens digits: $9 + 4 + 8 + 0 = 21$.

Step 3. Add the hundreds digits: $7 + 2 + 3 + 5 = 17$.

Step 4. Add the partial sums, as shown below.

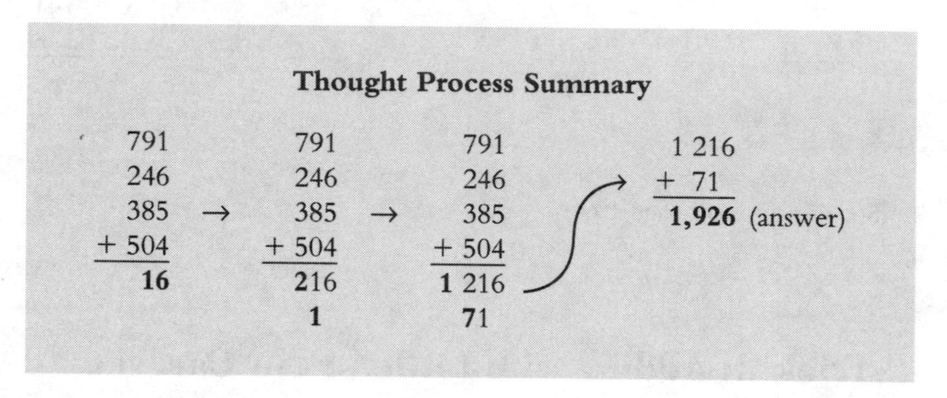

Brain Builder

4,903 + 8,665 + 5,122 + 7,835 + 6,019 + 1,984

Step 1. Add the ones digits: $3 + 5 + 2 + 5 + 9 + 4 = 28$.

Step 2. Add the tens digits: $0 + 6 + 2 + 3 + 1 + 8 = 20$.

Step 3. Add the hundreds digits: $9 + 6 + 1 + 8 + 0 + 9 = 33$.

Step 4. Add the thousands digits: $4 + 8 + 5 + 7 + 6 + 1 = 31$.

Step 5. Add the partial sums, as shown below.

Thought Process Summary

```
  4,903        4,903        4,903        4,903          33 228
  8,665        8,665        8,665        8,665        → + 1 30
  5,122        5,122        5,122        5,122          34,528 (answer)
  7,835 →      7,835 →      7,835 →      7,835
  6,019        6,019        6,019        6,019
+ 1,984      + 1,984      + 1,984      + 1,984
     28          228        3 228       33 228
                   0           30         1 30
```

Food for Thought: Occasionally, you will have to carry when adding the partial sums, though it wasn't necessary in our examples above. Also, when a partial sum turns out to be a one-digit number, it's best to convert it into a two-digit number with a zero to the left (otherwise, the next partial sum might be placed in the wrong column). So 8 would be written as 08.

Practice Problems

Proper placement of the partial sums is critical in working these exercises.

Elementary Exercises

1.
```
  628
  419
  343
+ 505
-----
1 025
   87
-----
1,895
```

2.
```
  144
  822
  539
+ 760
-----
```

3.
```
  939
  415
  670
+ 268
-----
```

4.
```
  222
  794
  108
+ 533
-----
```

5.
```
  414
  682
  370
+ 523
-----
```

6.
```
  466
  999
  321
+ 510
-----
```

7.
```
  190
  725
  613
+ 884
-----
```

8.
```
  396
  711
  288
+ 434
-----
```

9.
```
  591
  337
  899
+ 426
-----
```

10.
```
  100
  989
  444
+ 623
-----
```

11.
```
  776
  402
  934
+ 655
-----
```

12.
```
  999
  777
  555
+ 333
-----
```

13.
```
  808
  342
  177
+ 654
-----
```

14.
```
  471
  811
  732
+ 665
-----
```

Brain Builders

1.	3.	5.	7.
7,146	4,198	8,848	4,321
8,923	3,337	5,192	5,778
5,052	2,655	7,640	9,046
8,664	7,208	7,075	3,527
1,319	1,996	5,369	1,653
+ 4,287	+ 8,083	+ 1,852	+ 8,926

2.	4.	6.	8.
8,902	1,776	1,652	2,538
3,345	3,149	7,904	7,175
6,177	3,065	5,339	9,638
4,052	5,227	8,266	1,877
1,916	9,144	4,317	3,446
+ 4,560	+ 3,920	+ 2,055	+ 9,032

(See solutions on page 199.)

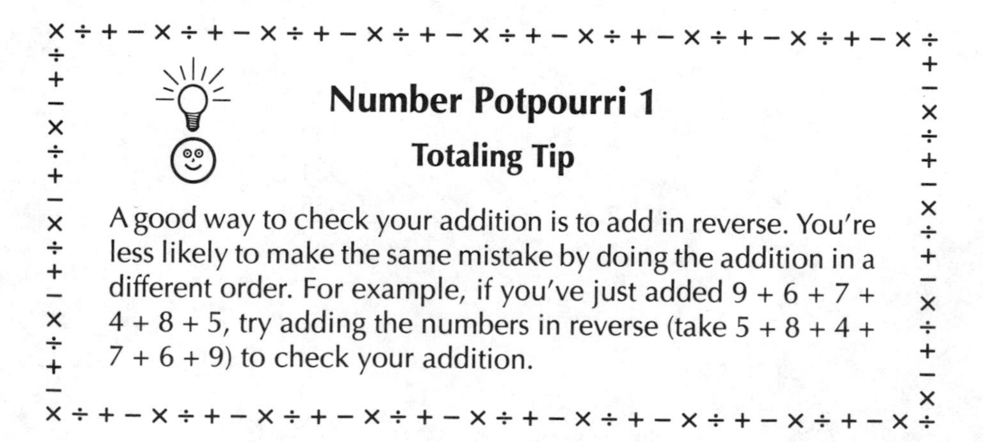

Number Potpourri 1

Totaling Tip

A good way to check your addition is to add in reverse. You're less likely to make the same mistake by doing the addition in a different order. For example, if you've just added 9 + 6 + 7 + 4 + 8 + 5, try adding the numbers in reverse (take 5 + 8 + 4 + 7 + 6 + 9) to check your addition.

Trick 3: Subtracting without Borrowing

Strategy: When performing a calculation, the only task worse than having to carry is having to borrow. Well, we solved the problem of carrying in Tricks 1 and 2. Now we're going to show you how to eliminate (in many cases) the need to borrow. Essentially, you are going to glance one column to the left to see if the calculation can be accomplished in one fell swoop. For example, in the calculation 573 − 254, forget about borrowing, and simply subtract 54 from 73 (and then subtract 2 from 5).

Elementary Example 1

762 − 437

Step 1. Notice that ones column subtraction cannot be done.

Step 2. Subtract ones and tens digits simultaneously: 62 − 37 = 25.

Step 3. Subtract hundreds digits: 7 − 4 = 3.

Thought Process Summary

$$\begin{array}{r} 762 \\ -\ 437 \\ \hline ? \end{array} \rightarrow \begin{array}{r} 762 \\ -\ 437 \\ \hline 25 \end{array} \rightarrow \begin{array}{r} 762 \\ -\ 437 \\ \hline 325 \end{array} \text{ (answer)}$$

Elementary Example 2

519 − 355

Step 1. Subtract ones digits: 9 − 5 = 4.

Step 2. Notice that tens column subtraction cannot be done.

Step 3. Subtract tens and hundreds digits simultaneously: 51 − 35 = 16.

Thought Process Summary

$$\begin{array}{r} 519 \\ -\,355 \\ \hline 4 \end{array} \rightarrow \begin{array}{r} 519 \\ -\,355 \\ \hline ?4 \end{array} \rightarrow \begin{array}{r} \mathbf{519} \\ -\,\mathbf{355} \\ \hline \mathbf{164} \end{array} \text{ (answer)}$$

Brain Builder

72,863 − 45,307

Step 1. Subtract ones and tens digits simultaneously: 63 − 07 = 56.

Step 2. Subtract hundreds digits: 8 − 3 = 5.

Step 3. Subtract thousands and ten thousands digits simultaneously: 72 − 45 = 27.

Thought Process Summary

$$\begin{array}{r} 72{,}8\mathbf{63} \\ -\,45{,}3\mathbf{07} \\ \hline \mathbf{56} \end{array} \rightarrow \begin{array}{r} 72{,}\mathbf{8}63 \\ -\,45{,}\mathbf{3}07 \\ \hline \mathbf{5}56 \end{array} \rightarrow \begin{array}{r} \mathbf{72},863 \\ -\,\mathbf{45},307 \\ \hline \mathbf{27},556 \end{array} \text{ (answer)}$$

Food for Thought: The success of this trick depends, in part, upon your ability to perform simultaneous two-digit subtraction. The best approach is to pretend you are adding. For instance, in Elementary Example 1, say to yourself, "37 plus what equals 62?" With practice, you'll be able to use this technique of subtracting by adding to more easily subtract without borrowing.

Practice Problems

Don't borrow when working these exercises. Instead, subtract in pairs.

Elementary Exercises

1. $\begin{array}{r} 471 \\ -\,357 \\ \hline \mathbf{114} \end{array}$

3. $\begin{array}{r} 629 \\ -\,477 \\ \hline \end{array}$

5. $\begin{array}{r} 963 \\ -\,458 \\ \hline \end{array}$

7. $\begin{array}{r} 370 \\ -\,145 \\ \hline \end{array}$

2. $\begin{array}{r} 843 \\ -\,516 \\ \hline \end{array}$

4. $\begin{array}{r} 768 \\ -\,491 \\ \hline \end{array}$

6. $\begin{array}{r} 646 \\ -\,483 \\ \hline \end{array}$

8. $\begin{array}{r} 507 \\ -\,181 \\ \hline \end{array}$

9. 949 − 657

10. 481 − 427

11. 873 − 258

12. 726 − 363

13. 318 − 127

14. 633 − 226

Brain Builders

1. 7,872 − 4,856

2. 4,916 − 3,674

3. 5,389 − 2,941

4. 8,507 − 1,351

5. 20,851 − 15,337

6. 52,509 − 36,259

7. 79,163 − 46,928

8. 42,500 − 18,375

(See solutions on page 199.)

Number Potpourri 2

Subtraction Check

When you're subtracting one number from another, you can check your answer really quickly by working backward. Let's say you just subtracted 138 from 414 and came up with the answer 276. At this point, simply take the 276 and add 138 to see if you get 414. This method works especially well when you're subtracting large numbers on your calculator.

Trick 4: Dividing with Even Numbers

Strategy: This trick is remarkably easy to apply. Whenever both numbers in a division problem are even, you can simplify the calculation by first cutting each number in half. If two even numbers result, cut those in half as well. When one of the numbers becomes odd, then solve the problem! That's all there is to it, as you'll see with our sample problems below.

Elementary Example 1

126 ÷ 14

Step 1. Cut 126 in half: 126 ÷ 2 = 63.

Step 2. Cut 14 in half: 14 ÷ 2 = 7.

Step 3. Divide 63 by 7: 63 ÷ 7 = 9 (answer).

> **Thought Process Summary**
>
> 126 ÷ 14 = 63 ÷ 7 = 9 (answer)

Elementary Example 2

154 ÷ 22

Step 1. Cut 154 in half: 154 ÷ 2 = 77.

Step 2. Cut 22 in half: 22 ÷ 2 = 11.

Step 3. Divide 77 by 11: 77 ÷ 11 = 7 (answer).

> **Thought Process Summary**
>
> 154 ÷ 22 = 77 ÷ 11 = 7 (answer)

Brain Builder

256 ÷ 32

Step 1. Cut 256 in half: 256 ÷ 2 = 128.

Step 2. Cut 32 in half: 32 ÷ 2 = 16. (You now have 128 ÷ 16, so you can cut those numbers in half.)

Step 3. Cut 128 in half: 128 ÷ 2 = 64.

Step 4. Cut 16 in half: 16 ÷ 2 = 8.

Step 5. Divide 64 by 8: 64 ÷ 8 = 8 (answer).

Thought Process Summary

$256 \div 32 = 128 \div 16 = 64 \div 8 = 8$ (answer)

Food for Thought: As is true with most other rapid math tricks, this one won't *always* produce a happy ending (as in our examples above). However, when working with even numbers, you can at least simplify the computation a little. Let's take the calculation $102 \div 12$. When reduced to $51 \div 6$, you can more easily determine that the answer is $8\frac{1}{2}$. (Actually, you could divide the 51 and the 6 by 3, to produce the still simpler calculation $17 \div 2$!)

Practice Problems

"Halve" yourself a ball with these practice exercises.

Elementary Exercises

1. $110 \div 22 =$ **$55 \div 11 = 5$**
2. $112 \div 14 =$
3. $144 \div 18 =$
4. $104 \div 26 =$
5. $144 \div 24 =$
6. $112 \div 16 =$
7. $140 \div 28 =$
8. $180 \div 36 =$
9. $132 \div 12 =$
10. $224 \div 32 =$
11. $176 \div 22 =$
12. $154 \div 14 =$
13. $114 \div 38 =$
14. $192 \div 16 =$

Brain Builders

1. $324 \div 36 =$
2. $242 \div 22 =$
3. $360 \div 24 =$
4. $416 \div 32 =$
5. $224 \div 16 =$
6. $576 \div 48 =$
7. $196 \div 28 =$
8. $352 \div 44 =$

(See solutions on page 200.)

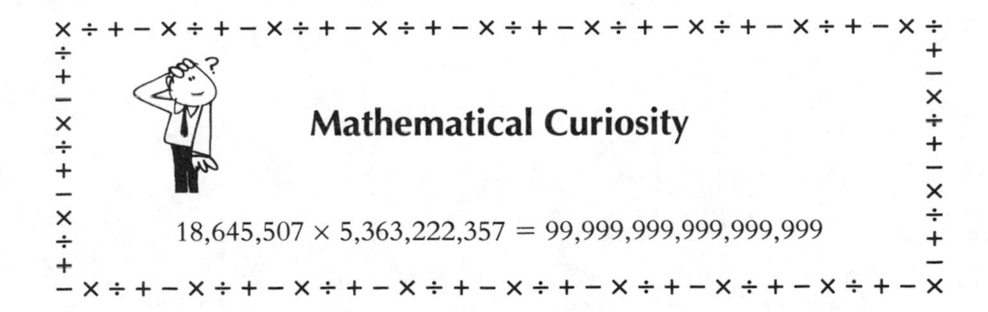

Trick 5: Adding Pluses and Minuses

Strategy: Every now and then one is faced with adding and subtracting a column of numbers. An example of such a calculation is 75 + 42 − 37 + 66 + 51 − 77. There are essentially three different ways to perform this type of calculation, as illustrated below.

Method 1: Perform the calculation, one number at a time.

75	
+ 42	(75 + 42 = 117)
− 37	(117 − 37 = 80)
+ 66	(80 + 66 = 146)
+ 51	(146 + 51 = 197)
− 77	(197 − 77 = 120, the answer)

Method 2: Add the pluses, then the minuses, and subtract.

	Pluses		Minuses		
75					
+ 42	75				
− 37	42				
+ 66	66		37		
+ 51	+ 51		+ 77		
− 77	234	−	114	=	120 (answer)

Method 3: Add, netting out the minuses against the pluses.

```
  75
+ 42 \
      > +5            75
- 37 /              +  5
              →
+ 66 \              - 11 \
+ 51  > -11                > +40
- 77 /              + 51 /
                    ----
                     120 (answer)
```

Food for Thought: The three methods have been presented in increasing order of efficiency. Though our third method may seem the most complicated, it's really the easiest method, once you become accustomed to it. Note also in Method 3 that we skipped ahead when netting. We then netted again. Obviously, there may be several different ways to net one number against another within a given calculation. In any case, use whichever method of the three you are most comfortable with.

Practice Problems

Try using either Method 2 or 3 for these exercises.

Elementary Exercises

```
1.   56 > 30        3.   44        5.   26        7.   88
   - 26                - 36           - 46           - 66
   - 63 > -18          + 59           + 63           + 17
   + 45        > -6    - 17           - 37           - 21
   + 39 > +12          + 50           + 51           + 43
   - 27                - 22           - 19           - 35
     24

2.   61             4.   79        6.   54        8.   14
   + 25                + 11           + 17           + 53
   - 47                - 66           - 25           - 49
   - 13                - 38           + 46           - 27
   + 54                + 43           - 33           + 64
   - 38                + 26           - 12           - 11
```

9. 28 + 45 − 15 + 37 − 31 − 52

10. 30 + 59 − 28 − 36 + 47 − 16

11. 49 − 36 + 27 − 13 + 48 − 25

12. 90 − 17 + 37 − 64 − 21 + 43

13. 33 + 18 − 24 + 59 − 49 − 15

14. 56 − 28 + 38 − 14 + 22 − 30

Brain Builders

1. 97 + 42 − 73 − 66 + 51 + 18

2. 71 − 54 + 33 + 19 − 46 − 23

3. 85 + 77 − 62 − 49 + 35 − 16

4. 62 − 36 − 18 + 45 + 95 − 51

5. 18 + 95 − 76 − 42 + 50 + 29

6. 64 − 42 + 79 − 57 + 33 − 18

7. 50 + 58 − 48 + 71 − 39 − 26

8. 99 − 64 − 35 + 56 + 19 − 42

(See solutions on page 200.)

Trick 6: Adding by Picking a Point of Reference

Strategy: When adding several numbers that are very close to each other, it's often faster to first guess at a midpoint and to then compute distances from that midpoint for each number. It involves a bit of simple multiplication as well. Let's see how this very practical trick works.

Elementary Example 1

		Distance from 80	
	78	− 2	
We'll	83	+ 3	(Did you use Trick 5
guess	81	+ 1	to combine these
at 80	77	− 3	pluses and minuses?)
as the	79	− 1	
midpoint.	82	+ 2	
	85	+ 5	
	+ 76	− 4	
(8 numbers ×	80 = 640)	+ 1	= 641 (answer)

Elementary Example 2

		Distance from 25
	28	+ 3
We'll	23	− 2
guess	30	+ 5
at 25	22	− 3
as the	24	− 1
midpoint.	25	0
	+ 20	− 5
(7 numbers ×	25 = 175)	− 3 = 172 (answer)

Brain Builder

		Distance from 200
	194	−6
We'll	207	+ 7
guess	201	+ 1
at 200	193	− 7
as the	206	+ 6
midpoint.	198	− 2
	202	+ 2
	197	− 3
	+ 204	+ 4

(9 numbers × 200 = 1,800) + 2 = 1,802 (answer)

Food for Thought: You're probably wondering, "What if my guess at the midpoint isn't very good?" Well, all that means is that your net distance will be a larger number to add or subtract. You should still obtain the correct answer. Actually, if in doubt about the approximate midpoint, it's best to guess at a nice, round number. If we had guessed at 199 in our brain builder example above, it would have been far more difficult to multiply 9 by 199 than to multiply 9 by 200.

Practice Problems

Make your best guess at the midpoint, and go from there!

Elementary Exercises

(Guess at 20)

1.
18	**− 2**
22	**+ 2**
23	**+ 3**
21	**+ 1**
19	**− 1**
17	**− 3**
20	**0**
+ 22	**+ 2**

(8 × 20) + 2 = 162

2.
13
14
17
16
14
13
18
+ 16

3.
43
36
41
37
39
42
38
+ 41

4.
28
23
23
24
27
25
22
+ 26

5.
9
12
11
8
13
7
9
+ 12

6.
63
57
59
62
60
58
61
+ 59

7.
28
27
33
29
30
32
31
+ 26

8.
82
82
79
80
83
76
78
+ 83

9.
49
52
51
47
50
48
53
+ 48

10.
92
87
90
89
93
91
88
+ 88

11.
67
72
71
69
69
73
66
+ 73

12.
21
22
23
18
17
19
20
+ 21

13.
36
37
34
33
36
36
32
+ 38

14.
61
61
59
57
63
62
58
+ 63

Brain Builders

1.	3.	5.	7.
103	147	406	247
98	152	404	259
95	150	395	251
105	148	403	244
97	146	392	245
101	153	398	257
+ 104	+ 152	+ 402	+ 249

2.	4.	6.	8.
596	892	691	1,012
601	897	704	995
605	904	710	997
593	903	696	1,002
607	895	698	1,008
599	899	705	998
+ 598	+ 906	+ 702	+ 989

(See solutions on page 200.)

Number Potpourri 3

Sum Fun

1	1
1 2	2 1
1 2 3	3 2 1
1 2 3 4	4 3 2 1
1 2 3 4 5	5 4 3 2 1
1 2 3 4 5 6	6 5 4 3 2 1
1 2 3 4 5 6 7	7 6 5 4 3 2 1
1 2 3 4 5 6 7 8	8 7 6 5 4 3 2 1
+ 1 2 3 4 5 6 7 8 9	+ 9 8 7 6 5 4 3 2 1
1,0 8 3,6 7 6,2 6 9	1,0 8 3,6 7 6,2 6 9

Trick 7: Multiplying with Factors

Strategy: This trick is best applied when multiplying by a two-digit number that can be expressed as two one-digit factors. For example, the calculation 346×28 can be solved more efficiently as $346 \times 7 \times 4$ (or as $346 \times 4 \times 7$). You'll note in the following examples that the addition of partial sums (that takes place with routine multiplication) has been eliminated.

Elementary Example 1

714 × 27

Step 1. Determine two one-digit factors of 27: 9 and 3.

Step 2. Multiply 714 by 9, then by 3 (or vice versa).

Thought Process Summary

$$\begin{array}{r} 714 \\ \times \quad 9 \\ \hline 6{,}426 \\ \times \quad 3 \\ \hline 19{,}278 \end{array} \text{ (answer)}$$

Elementary Example 2

293 × 35

Step 1. Determine two one-digit factors of 35: 7 and 5.

Step 2. Multiply 293 by 7, then by 5 (or vice versa).

Thought Process Summary

$$\begin{array}{r} 293 \\ \times \quad 7 \\ \hline 2{,}051 \\ \times \quad 5 \\ \hline 10{,}255 \end{array} \text{ (answer)}$$

Brain Builder

3,506 × 48

Step 1. Determine two one-digit factors of 48: 8 and 6.

Step 2. Multiply 3,506 by 8, then by 6 (or vice versa).

Thought Process Summary

$$\begin{array}{r} 3{,}506 \\ \times \quad 8 \\ \hline 28{,}048 \\ \times \quad 6 \\ \hline 168{,}288 \end{array} \text{ (answer)}$$

Food for Thought: A few two-digit numbers can be expressed as more than one pair of one-digit factors. For example, you could multiply by 36 by multiplying by 9, then by 4 (or vice versa). Instead, however, you could multiply by 6, then by 6 again.

Practice Problems

Convert the multiplier into factors, then multiply.

Elementary Exercises

1. 144 × 15 =

$$\begin{array}{r} \times \quad \mathbf{5} \\ \hline \mathbf{720} \\ \times \quad \mathbf{3} \\ \hline \mathbf{2{,}160} \end{array}$$

2. 327 × 49 =

3. 269 × 28 =

4. $734 \times 56 =$
5. $518 \times 72 =$
6. $283 \times 42 =$
7. $469 \times 32 =$
8. $637 \times 27 =$
9. $195 \times 64 =$
10. $373 \times 54 =$
11. $747 \times 24 =$
12. $582 \times 48 =$
13. $855 \times 36 =$
14. $499 \times 25 =$

Brain Builders

1. $2{,}486 \times 35 =$
2. $5{,}133 \times 18 =$
3. $7{,}448 \times 45 =$
4. $4{,}917 \times 63 =$
5. $1{,}928 \times 21 =$
6. $6{,}844 \times 36 =$
7. $3{,}298 \times 72 =$
8. $9{,}999 \times 56 =$

(See solutions on page 201.)

Trick 8: Multiplying Out of Order

Strategy: Now and then you must multiply more than two numbers together, as when you are obtaining a cubic measure. As it's often easier to add numbers out of order, multiplying numbers out of order can also make calculating much simpler. Let's try this technique on some unsuspecting numbers.

Elementary Example 1

$4 \times 17 \times 25$

Step 1. Recast the computation as $4 \times 25 \times 17$.

Step 2. Multiply the first two numbers, then the third.

Thought Process Summary

$4 \times 17 \times 25 = (4 \times 25) \times 17 = 100 \times 17 = 1{,}700$ (answer)

Elementary Example 2

$25 \times 13 \times 8$

Step 1. Recast the computation as $25 \times 8 \times 13$.

Step 2. Multiply the first two numbers, then the third.

Thought Process Summary

$25 \times 13 \times 8 = (25 \times 8) \times 13 = 200 \times 13 = 2{,}600$ (answer)

Brain Builder

$60 \times 32 \times \frac{1}{3}$

Step 1. Recast the computation as $60 \times \frac{1}{3} \times 32$.

Step 2. Multiply the first two numbers, then the third.

Thought Process Summary

$60 \times 32 \times \frac{1}{3} = (60 \times \frac{1}{3}) \times 32 = 20 \times 32 = 640$ (answer)

Food for Thought: A related trick, which combines Tricks 7 and 8, is to first break apart one of the factors and then multiply. For example, the calculation 25 × 24 could be viewed as 25 × (4 × 6), which could be rewritten more simply as (25 × 4) × 6, which equals 100 × 6, or 600.

Practice Problems

Multiply the numbers in whatever order is easiest for you.

Elementary Exercises

1. 2 × 7 × 25 = **(2 × 25) × 7 = 350**
2. 15 × 9 × 4 =
3. 6 × 11 × 5 =
4. 35 × 7 × 2 =
5. 8 × 9 × 25 =
6. 5 × 12 × 8 =
7. 2 × 7 × 45 =
8. 2 × 8 × 3 × 5 =
9. 25 × 3 × 7 × 4 =
10. 2 × 4 × 3 × 15 =
11. 5 × 3 × 8 × 5 =
12. 6 × 2 × 5 × 3 =
13. 25 × 3 × 6 × 2 =
14. 65 × 5 × 2 × 2 =

Brain Builders

1. $11 \times \frac{1}{2} \times 14 =$
2. $24 \times 9 \times \frac{1}{4} =$
3. $\frac{1}{3} \times 17 \times 30 =$
4. $35 \times 7 \times \frac{1}{5} =$
5. $48 \times 5 \times \frac{1}{4} \times 2 =$
6. $\frac{1}{7} \times 5 \times 56 \times 4 =$
7. $\frac{1}{3} \times 22 \times 15 \times \frac{1}{2} =$
8. $45 \times \frac{1}{4} \times 28 \times \frac{1}{5} =$

(See solutions on page 201.)

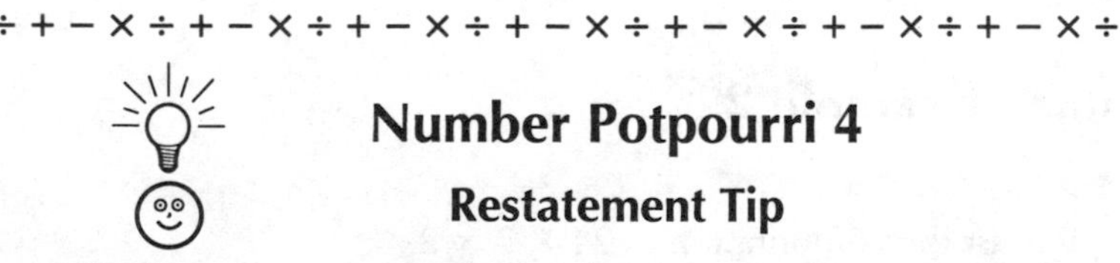

Number Potpourri 4

Restatement Tip

Sometimes a number presented in decimal form is more easily manipulated when rewritten in fraction form. The reverse can be true, as well. For example, wouldn't the calculation 28 × 0.25 be easier to solve as $28 \times \frac{1}{4}$? Similarly, wouldn't $12 \div \frac{2}{5}$ be far less imposing if approached as 12 ÷ 0.4? That's why you should have the important decimal/fraction equivalents committed to memory. In fact, they can be found in this book on page 8!

Trick 9: Multiplying by 6

Strategy: The smaller the number, the more likely it will be involved in a computation. The number 6 is pretty small and thus is commonly calculated upon. Because 6 can be expressed as 3×2 (or as 2×3), a calculation such as 19×6 can be more easily solved if viewed as $19 \times 3 \times 2$. Let's take a closer look at this very basic rapid math trick.

Elementary Example 1

17 × 6

Step 1. Recast the computation as $17 \times 3 \times 2$.

Step 2. Multiply the first two numbers, then the third.

Thought Process Summary

$17 \times 6 = 17 \times (3 \times 2) = (17 \times 3) \times 2 = 51 \times 2 = 102$ (answer)

Elementary Example 2

24 × 6

Step 1. Recast the computation as $24 \times 3 \times 2$.

Step 2. Multiply the first two numbers, then the third.

Thought Process Summary

$24 \times 6 = 24 \times (3 \times 2) = (24 \times 3) \times 2 = 72 \times 2 = 144$ (answer)

Brain Builder

45 × 6

Step 1. Recast the computation as 45 × 2 × 3.
Step 2. Multiply the first two numbers, then the third.

Thought Process Summary

$45 \times 6 = 45 \times (2 \times 3) = (45 \times 2) \times 3 = 90 \times 3 = 270$ (answer)

Food for Thought: As you learned in Trick 7, many numbers can be split into factors to make a computation simpler. However, splitting the number 6 into 3 × 2 (or 2 × 3) works very well because 3 and 2 are such tiny, easy numbers to work with. One final note — in our Brain Builder example above, we could have instead recast the computation as 45 × 3 × 2. However, the calculation would have been a wee bit more difficult to execute.

Practice Problems

View 6 as 3 × 2 or 2 × 3 when working these exercises.

Elementary Exercises

1. 18 × 6 = **(18 × 3) × 2 = 108**
2. 22 × 6 =
3. 13 × 6 =
4. 32 × 6 =
5. 27 × 6 =
6. 16 × 6 =
7. 23 × 6 =
8. 15 × 6 =
9. 19 × 6 =
10. 35 × 6 =
11. 33 × 6 =
12. 17 × 6 =
13. 24 × 6 =
14. 14 × 6 =

Brain Builders

1. 55 × 6 =
2. 34 × 6 =
3. 26 × 6 =
4. 85 × 6 =
5. 36 × 6 =
6. 42 × 6 =
7. 65 × 6 =
8. 37 × 6 =

(See solutions on page 201.)

Trick 10: Dividing by 6

Strategy: As you have probably guessed, this trick works in much the same way as Trick 9. That is, we are going to divide by 6 by first dividing by 2, then 3, or by first dividing by 3, then 2. For example, $108 \div 6$ can be solved far more easily by first taking $108 \div 2 = 54$. Then take $54 \div 3$, and you've got your answer, 18.

Elementary Example 1

150 ÷ 6

Step 1. Recast the computation as $150 \div 3 \div 2$.

Step 2. Divide: $150 \div 3 = 50$.

Step 3. Divide: $50 \div 2 = 25$ (answer).

Thought Process Summary

$$150 \div 6 = (150 \div 3) \div 2 = 50 \div 2 = 25 \text{ (answer)}$$

Elementary Example 2

132 ÷ 6

Step 1. Recast the computation as $132 \div 2 \div 3$.

Step 2. Divide: $132 \div 2 = 66$.

Step 3. Divide: $66 \div 3 = 22$ (answer).

Thought Process Summary

$$132 \div 6 = (132 \div 2) \div 3 = 66 \div 3 = 22 \text{ (answer)}$$

Brain Builder

270 ÷ 6

Step 1. Recast the computation as $270 \div 3 \div 2$.

Step 2. Divide: $270 \div 3 = 90$.

Step 3. Divide: $90 \div 2 = 45$ (answer).

Thought Process Summary

$$270 \div 6 = (270 \div 3) \div 2 = 90 \div 2 = 45 \text{ (answer)}$$

Food for Thought: Perhaps you have figured out that each of the three problems presented above could have been solved by reversing the factors of 6. For example, our Brain Builder could have been solved instead as $(270 \div 2) \div 3$. In each case, you must make a judgment call regarding which digit (2 or 3) to divide by first.

Practice Problems

Divide by 3, and then by 2 (or vice versa) to simplify division by 6.

Elementary Exercises

1. $210 \div 6 = \mathbf{(210 \div 3) \div 2 = 35}$
2. $90 \div 6 =$
3. $108 \div 6 =$
4. $192 \div 6 =$
5. $102 \div 6 =$
6. $138 \div 6 =$
7. $114 \div 6 =$
8. $96 \div 6 =$
9. $144 \div 6 =$
10. $78 \div 6 =$
11. $198 \div 6 =$
12. $84 \div 6 =$
13. $162 \div 6 =$
14. $132 \div 6 =$

Brain Builders

1. $330 \div 6 =$
2. $780 \div 6 =$
3. $450 \div 6 =$
4. $216 \div 6 =$
5. $510 \div 6 =$
6. $204 \div 6 =$
7. $840 \div 6 =$
8. $390 \div 6 =$

(See solutions on page 202.)

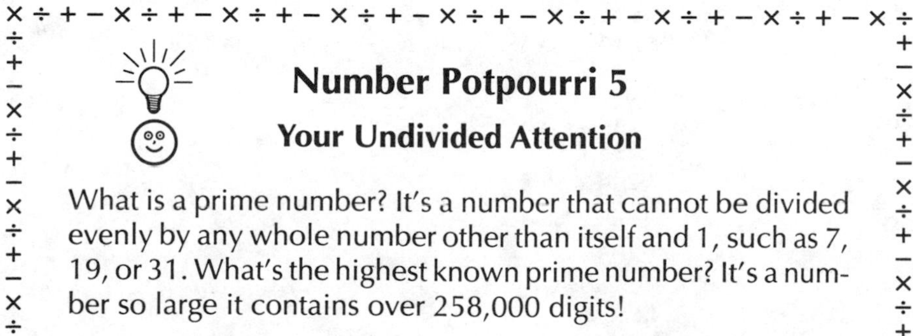

Number Potpourri 5

Your Undivided Attention

What is a prime number? It's a number that cannot be divided evenly by any whole number other than itself and 1, such as 7, 19, or 31. What's the highest known prime number? It's a number so large it contains over 258,000 digits!

Trick 11: Adding by Multiplying

Strategy: This trick works best when you are adding a long column of numbers and one or two of the digits are repeated several times. In such cases, it's best to count how many times the digit is contained in a column, multiply, and then add. Let's take a look at an example or three.

Elementary Example 1

```
   3
  43
  47
  51     Tens column repetition
  46         Six 4's = 24
  45         Two 5's = 10
  50         Carried =  3
  49                   --
+ 42                   37
----
 373
```

Elementary Example 2

```
   3
  25
  35
  70     Ones column repetition
  65        Seven 5's = 35
  45     (write the 5, carry the 3)
  25
  85
+ 55
----
 405
```

Brain Builder

5 4	Tens column repetition
298	Six 9's = 54
395	Carried = 4
202	58 (write the 8, carry the 5)
299	
291	Hundreds column repetition
305	Five 2's = 10
296	Three 3's = 9
+ 397	Carried = 5
2,483	24

Food for Thought: You can apply this technique to the simpler computations (involving fewer or smaller numbers) by performing the multiplication and addition in your head (as you could in Elementary Example 2 above). However, when you are adding very long columns of numbers, it may be necessary to do some "side calculations" on scratch paper, as we have illustrated. Finally, for those who know how to add from left to right, remember that (for example) a 7 in the tens column represents 70, not 7.

Practice Problems

Watch for repeating digits when solving these.

Elementary Exercises

1.			3.	5.	7.
	35		63	93	41
	75		47	91	38
	14	**(Multiply**	73	94	43
	65	**six 5's**	93	77	49
	85	**in ones**	13	90	46
	92	**column.)**	97	98	50
	15		53	92	44
	+ 25		+ 23	+ 86	+ 27
	406				

2.	4.	6.	8.
81	24	36	87
82	26	56	47
86	52	76	47
84	25	26	66
85	25	16	77
88	28	96	17
89	21	66	57
+ 84	+ 27	+ 46	+ 37

9.	11.	13.
52	29	60
57	59	57
35	19	64
51	40	61
59	89	69
50	39	53
56	49	65
+ 54	+ 31	+ 58

10.	12.	14.
68	32	32
35	92	75
32	72	75
34	52	68
67	12	75
31	82	75
65	42	75
+ 33	+ 62	+ 75

Brain Builders

1.	3.	5.	7.
374	798	902	437
164	314	951	427
704	366	932	537
629	317	962	431
514	355	612	437
964	323	972	537
+ 224	+ 102	+ 943	+ 421

2.	4.	6.	8.
671	513	368	709
413	519	168	739
874	916	463	769
272	518	468	549
975	537	728	703
306	514	260	729
+ 770	+ 511	+ 868	+ 589

(See solutions on page 202.)

Trick 12: Adding Amounts Just under Multiples of $1

Strategy: This trick works best with calculations such as $1.98 + $4.93 + $2.95. The secret is to round each amount to the nearest whole dollar, add, and then subtract the sum of the amounts added during the rounding process. Let's look at some samples of this easy-to-apply technique.

Elementary Example 1

Actual amount	Rounded	Add-on amount
$2.95	$3	5¢
4.90	5	10
1.97	2	3
3.92	4	8
+ 5.96	6	4
	$20 −	30¢ = $19.70 (answer)

Elementary Example 2

Actual amount	Rounded	Add-on amount
$3.93	$4	7¢
6.96	7	4
1.92	2	8
4.97	5	3
+ 2.91	3	9
	$21 −	31¢ = $20.69 (answer)

Brain Builder

Actual amount	Rounded	Add-on amount
$7.94	$8	6¢
5.88	6	12
3.98	4	2
8.85	9	15
6.90	7	10
+ 9.97	10	3
	$44 −	48¢ = $43.52 (answer)

Food for Thought: Try this trick at the supermarket (though most likely you ordinarily require only an estimate of the cost of groceries purchased). The amounts, incidentally, do not have to be in dollar-and-cents form for this trick to work. You could add 298 + 794 + 597, and so on, by just rounding to the nearest hundred, adding, and then subtracting the amounts added on during the rounding process. Although this trick is very easy to apply, many people don't think of using it.

Practice Problems

Remember to round up, add, then deduct the add-on amounts.

Elementary Exercises

1.

$2.95	=	**$3 − 5¢**
4.92	=	**5 − 8**
5.97	=	**6 − 3**
3.98	=	**4 − 2**
+ 1.96	=	**2 − 4**
$19.78	=	**$20 − 22¢**

2.

$6.90
3.98
1.92
2.99
+ 4.97

3.

$3.92
2.97
5.96
6.91
+ 1.99

4.

$4.95
4.95
1.91
6.98
+ 3.94

5.

$5.94
1.98
4.97
6.97
+ 3.90

6.

$1.93
6.90
5.91
4.97
+ 6.95

7.

$4.98
6.99
3.97
5.96
+ 1.98

8.

$2.92
4.90
6.91
5.93
+ 3.92

9.

$2.92
4.95
6.98
1.97
+ 3.93

10. $4.94
1.96
3.91
6.90
+ 5.99

11. $5.95
1.92
3.96
4.98
+ 5.91

12. $3.98
6.97
1.99
2.96
+ 4.98

13. $4.90
6.97
5.94
3.95
+ 2.98

14. $1.97
6.91
5.95
4.93
+ 2.99

Brain Builders

1. $8.88
4.95
7.96
9.85
1.92
+ 6.89

2. $3.98
9.86
7.99
5.83
6.95
+ 2.89

3. $6.85
9.94
1.92
8.98
5.87
+ 7.80

4. $7.96
9.89
2.95
4.91
6.85
+ 1.94

5. $1.84
7.98
5.93
8.89
6.99
+ 3.95

6. $4.95
9.88
1.96
5.92
3.80
+ 7.97

7. $6.85
7.92
1.93
5.86
8.95
+ 9.99

8. $8.98
1.83
7.95
4.91
9.88
+ 3.96

(See solutions on page 202.)

Trick 13: Multiplying with Multiples of Numbers

Strategy: This trick will work when the multiplier contains digits that are multiples of each other. The number 62 qualifies, because the tens digit (6) is a multiple of the ones digit (2). When multiplying by 62, we would first multiply by 2. Then, instead of multiplying by 6, we would triple the partial product obtained from multiplying by 2, remembering to move one place to the left. This technique would also work for a number such as 26, though we would have to take one-third of the first partial product to obtain the second one, since 2 is one-third of 6. The theory behind this trick is that multiplying (or dividing) is easiest when it involves tiny numbers such as 2 or 3.

Elementary Example 1

72 × 63

Step 1. Multiply 72 by 3: 72 × 3 = 216.

Step 2. Double the 216, since 6 is the double of 3, placing the result one column to the left.

Step 3. Add, and you've got the answer.

Thought Process Summary

```
   72
 × 63
  216 ← Product of 72 × 3.
 4 32 ← Double the 216, because 6 is the double of 3.
4,536 ← Add the partial products to produce the answer.
```

Elementary Example 2

37 × 24

Step 1. Multiply 37 by 4: 37 × 4 = 148.

Step 2. Halve the 148, since 2 is half of 4, placing the result one column to the left.

Step 3. Add, and you've got the answer.

Thought Process Summary

```
   37
 × 24
  148   ←  Product of 37 × 4.
  74    ←  Halve the 148, because 2 is half of 4.
  888   ←  Add the partial products to produce the answer.
```

Brain Builder

24 × 126

Step 1. Multiply 24 by 6: 24 × 6 = 144.

Step 2. Double the 144, since 12 is double 6, placing the result one column to the left. (Notice that the tens and hundreds digits are viewed together as 12.)

Step 3. Add, and you've got the answer.

Thought Process Summary

```
    24
 × 126
   144   ←  Product of 24 × 6.
  2 88   ←  Double the 144, since 12 is the double of 6.
 3,024   ←  Add the partial products to produce the answer.
```

Food for Thought: Elementary Example 2 could have been solved in reverse, as follows.

```
   37
 × 24
  74  ← Product of 37 × 2, written one column to left.
 148  ← Double the 74, because 4 is the double of 2.
 888  ← Add the partial products to produce the answer.
```

Note the placement of the 148. Remember that it is the product of 37 and the ones digit of 24.

Practice Problems

Check for "digit-multiples" when working these exercises.

Elementary Exercises

```
1.    65        5.    33        9.    45       13.    18
    × 42            × 63            × 41              × 82
     130
    2 60  (130 × 2)
   2,730

2.    25        6.    18       10.    16       14.    65
    × 31            × 84            × 62              × 24

3.    36        7.    21       11.    64
    × 21            × 93            × 42

4.    48        8.    76       12.    32
    × 12            × 36            × 48
```

Brain Builders

1. 22×421
2. 25×147
3. 32×842
4. 48×105
5. 35×124
6. 15×189
7. 54×126
8. 88×510

(See solutions on page 203.)

Parlor Trick 1: The Mystifying Missing-Digit Trick

There are several different ways to perform this trick. Here is the most impressive way.

First, make sure you have a calculator handy. Then ask someone (your subject) to write down any whole number, such as 496 or 7,355. It doesn't matter what size the number is, but you might suggest that it be limited to a four-digit number so that your calculator will be able to accommodate the computations to follow. You are not allowed to see the number, but make sure your subject shows it to your audience (assuming you have one). You can either turn your back or wear a blindfold (for special effect!).

Next, have your subject rewrite the number with the digits in a different order. So the number 7,355 could be rewritten as 5,753, or any other way that uses the digits 7, 3, 5, and 5.

Then have your subject subtract the smaller number from the larger one, using the calculator. So in our example, we would have 7,355 – 5,753 = 1,602 (assuming 7,355 is rearranged as 5,753).

The next step is to have your subject multiply the above result (1,602 in our case) by any number, using the calculator. You should probably limit it to a four-digit number, again to make sure your calculator can accommodate the result.

Let's say our subject chose the number 7,144 to multiply by 1,602. The product would be 11,444,688, which should be written in plain sight for everyone in the audience (except for you, of course).

Next, have your subject circle any one of the product's digits, specifying that it must be a digit from 1 through 9 and not zero (if you were to allow your subject to choose zero, the trick might not work the first time).

Finally, have your subject read aloud the remaining digits of the product (the ones not circled), slowly, and in any order. You might suggest that he or she cross out each digit as it is read, to avoid double-counting or omission. What you are going to do is reveal the identity of the missing digit (the one that is circled, and thus was not read aloud).

Here's the Trick: If your subject has executed each step correctly, all you have to do is add up the digits as he or she reads them aloud. The missing digit will then be the one that will produce the next multiple of 9. Multiples of 9 are 9, 18, 27, 36, and so on. In our example above (product = 11,444,688), let's say our subject has circled the digit 6. He or she then reads aloud all the other digits, in random order. They total 30. You then ask yourself, "30 plus what will produce the next multiple of 9." Since the next multiple of 9 (after 30) is 36, then the missing digit must be 36 – 30, or 6!

This trick is based on a concept known as "casting out nines." What is most amazing about the trick is that at no time do you know what any of the numbers are, yet you will be able to identify a randomly chosen missing digit. Even the digits of the product are read in random order. Nevertheless, the trick will always work when carefully executed by both parties.

If you'd like to simplify the overall procedure, simply ask your subjects to begin with the number 9,999 and have them multiply it by any other four-digit number. Have them circle a digit of the product (1 through 9, not zero), and go from there. It's truly an amazing trick that never fails to astound and impress.

Summary of Steps

1. Have your subject write down a number of up to four digits.
2. Have the subject rearrange the digits of the number.
3. Have the subject subtract the smaller number from the larger.
4. Have the subject multiply the result by any number, up to four digits.
5. Have the subject circle any digit of the product, from 1 to 9.
6. Have the subject read aloud, in random order, all the other digits.
7. You will add up the digits in your head; the missing (circled) digit will be the one that produces the next multiple of 9.

Week 1 Quick Quiz

Let's see how many tricks from Week 1 you can remember and apply by taking this brief test. There's no time limit, but try to work through these items as rapidly as possible. Before you begin, glance at the computations and try to identify the trick that you could use. In some instances, however, you will be asked to perform a calculation in a certain manner. When you flip ahead to the solutions, you will see which trick was intended.

Elementary Exercises

1. Multiply without carrying:
$$\begin{array}{r} 73 \\ \times\ 8 \\ \hline \end{array}$$

2. Add without carrying:
$$\begin{array}{r} 519 \\ 772 \\ 834 \\ +\ 366 \\ \hline \end{array}$$

3. $$\begin{array}{r} 533 \\ -\ 126 \\ \hline \end{array}$$

4. $224 \div 16 =$

5. $$\begin{array}{r} 56 \\ +\ 22 \\ -\ 17 \\ -\ 32 \\ +\ 28 \\ \hline \end{array}$$

6. $$\begin{array}{r} 32 \\ 29 \\ 27 \\ 33 \\ 28 \\ +\ 33 \\ \hline \end{array}$$

7. Multiply by factors:
$$\begin{array}{r} 327 \\ \times\ 56 \\ \hline \end{array}$$

8. $15 \times 5 \times 4 \times 2 =$

9. $17 \times 6 =$

10. $144 \div 6 =$

11. $$\begin{array}{r} 35 \\ 85 \\ 15 \\ 22 \\ 45 \\ +\ 75 \\ \hline \end{array}$$

12. $$\begin{array}{r} \$2.95 \\ 5.92 \\ 3.97 \\ 1.98 \\ +\ 4.91 \\ \hline \end{array}$$

13. $$\begin{array}{r} 35 \\ \times\ 62 \\ \hline \end{array}$$

Brain Builders

1. Multiply without carrying:

$$\begin{array}{r} 538 \\ \times\ 7 \\ \hline \end{array}$$

2. $$\begin{array}{r} 6,862 \\ -\ 3,646 \\ \hline \end{array}$$

3. $420 \div 28 =$

4. Multiply by factors:

$$\begin{array}{r} 2,185 \\ \times\ 36 \\ \hline \end{array}$$

5. $22 \times 18 \times \frac{1}{3} \times \frac{1}{2} =$

6. $55 \times 6 =$

7. $$\begin{array}{r} \$7.96 \\ 9.88 \\ 6.93 \\ 2.85 \\ +\ 8.95 \\ \hline \end{array}$$

(See solutions on page 218.)

Number Potpourri 6

The Dangers of Rounding

Rounding is one of the most powerful tools employed in the world of rapid calculation. However, we must be extremely careful when determining to what degree to round. For example, the circumference of the moon (6,786 miles) is calculated by multiplying its diameter (we'll assume exactly 2,160 miles) by π (3.14159265 . . .). Let's see how inaccurate our calculation would be given the following rounded values of π:

$2,160 \times 3$	= about 306 miles off
$2,160 \times 3.1$	= about 90 miles off
$2,160 \times 3.14$	= about $3\frac{1}{2}$ miles off
$2,160 \times 3.142$	= about $\frac{9}{10}$ of a mile off
$2,160 \times 3.1416$	= about 28 yards off

Week 2 More Intermediate Rapid Math Tricks

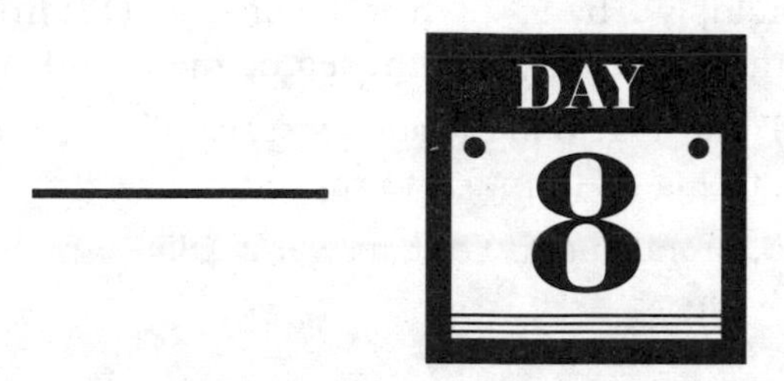

Trick 14: Dividing by Showing Little or No Work

Strategy: We begin Week 2 with a trick intended to conserve space and, hopefully, time. It involves subtracting in the conventional way except that much of the work is done in your head. The less work you have to write down, the faster you will be able to execute the trick. Let's see how this unconventional technique for dividing works.

Elementary Example 1

3,073 ÷ 7

Step 1. Divide 7 into 30, and write 4 in the answer space.

Step 2. In your head, multiply 4 by 7, subtract the product (28) from 30, and write only the difference (2) to the left of the next digit in the dividend (7).

Step 3. Divide 7 into 27 and write the 3 in the answer space. Place the difference of 6 (from taking 21 from 27) to the left of the 3 in the dividend.

Step 4. Divide 7 into 63, producing 9. Your answer is 439.

Thought Process Summary

$$\begin{array}{r} 4\ \ 3\ \ 9 \\ 7\overline{)\,3\,,\,0\ {}^{2}7\ {}^{6}3} \end{array}$$

Elementary Example 2

1,710 ÷ 6

Step 1. Divide 6 into 17, and write 2 in the answer space.

Step 2. In your head, multiply 2 by 6, subtract the product (12) from 17, and write only the difference (5) to the left of the next digit (1).

Step 3. Divide 6 into 51 and write 8 in the answer space. Place the difference of 3 (from taking 48 from 51) to the left of the zero.

Step 4. Divide 6 into 30, producing 5. Your answer is 285.

Thought Process Summary

$$\begin{array}{r} 2\ 8\ 5 \\ 6\overline{)1,7\,{}^{5}1\,{}^{3}0} \end{array}$$

Brain Builder

8,712 ÷ 11

Step 1. Divide 11 into 87, and write 7 in the answer space.

Step 2. In your head, multiply 7 by 11, subtract the product (77) from 87, and write only the difference (10) to the left of the next digit (1).

Step 3. Divide 11 into 101 and write 9 in the answer space. Place the difference of 2 (from taking 99 from 101) to the left of the 2.

Step 4. Divide 11 into 22, producing 2. Your answer is 792.

Thought Process Summary

$$\begin{array}{r} 7\ \ 9\ 2 \\ 11\overline{)8,7\,{}^{10}1\,{}^{2}2} \end{array}$$

Food for Thought: When we calculate the "long" way, so much time is normally taken writing down the work. This trick saves time by eliminating the majority of that work. In fact, with practice, you won't even have to write down the little remainders that are carried after each subtraction! (Why not try it?)

Practice Problems

Just multiply and subtract in your head, and you'll save time and space.

Elementary Exercises

1. $\begin{array}{r} \mathbf{3\ 4\ 7} \\ 8\overline{)2,7^{3}7^{5}6} \end{array}$
2. $5\overline{)2,445}$
3. $9\overline{)1,494}$
4. $7\overline{)1,764}$
5. $3\overline{)2,502}$
6. $6\overline{)3,714}$
7. $4\overline{)3,888}$
8. $9\overline{)4,572}$
9. $7\overline{)5,348}$
10. $5\overline{)1,940}$
11. $8\overline{)1,096}$
12. $4\overline{)3,996}$
13. $3\overline{)2,361}$
14. $9\overline{)2,322}$

Brain Builders

1. $12\overline{)2,928}$
2. $15\overline{)8,115}$
3. $11\overline{)10,692}$
4. $13\overline{)4,485}$
5. $22\overline{)9,152}$
6. $25\overline{)4,425}$
7. $16\overline{)5,696}$
8. $33\overline{)7,095}$

(See solutions on page 203.)

Trick 15: Dividing with Factors

Strategy: This trick is similar to Trick 7 in that you will be converting a two-digit factor into two one-digit factors to simplify the computation. For example, the calculation 1,785 ÷ 21 can be solved more efficiently as 1,785 ÷ 7 ÷ 3 (or as 1,785 ÷ 3 ÷ 7). Even though two divisions will be performed instead of one, many people are intimidated by two-digit divisors and would much prefer dealing with one digit only. You'll note in our examples below that we will incorporate the technique you just learned in Trick 14: dividing by showing little or no work.

Elementary Example 1

1,785 ÷ 21

Step 1. Determine two one-digit factors of 21: 7 and 3.

Step 2. Divide 1,785 by 7, showing little or no work.

Step 3. Divide the result of Step 2 by 3 to obtain your answer.

Thought Process Summary

$$\begin{array}{r} 8\ 5 \text{ (answer)} \\ 3\overline{)2\ 5\ {}^{1}5} \\ 7\overline{)1,7\ {}^{3}8\ {}^{3}5} \end{array}$$

Elementary Example 2

3,087 ÷ 49

Step 1. Determine two one-digit factors of 49: 7 and 7.

Step 2. Divide 3,087 by 7, showing little or no work.

Step 3. Divide the result of Step 2 by 7 to obtain your answer.

Thought Process Summary

$$\begin{array}{r} 6\ 3 \text{ (answer)} \\ 7\overline{)4\ 4\ {}^{2}1} \\ 7\overline{)3,0\ {}^{2}8\ 7} \end{array}$$

Brain Builder

25,632 ÷ 72

Step 1. Determine two one-digit factors of 72: 9 and 8.

Step 2. Divide 25,632 by 9, showing little or no work.

Step 3. Divide the result of Step 2 by 8 to obtain your answer.

Thought Process Summary

$$\begin{array}{r} 3\ 5\ 6 \text{ (answer)} \\ 8\overline{)\,2,8\,{}^{4}4\,{}^{4}8} \\ 9\overline{)\,2\ 5,{}^{7}6\,{}^{4}3\,{}^{7}2} \end{array}$$

Food for Thought: Although we've used Trick 14 to minimize the amount of work written down, Trick 15 will work just fine the conventional way. That is, just divide by one factor the usual way (showing all the work), and then divide the result by the other factor the usual way. In addition, you've probably already figured out this trick's one shortcoming — things might not work out evenly, as they did in the sample problems above. In those cases, however, you would probably have to deal with remainders or decimals using the conventional way anyway.

Practice Problems

Simplify these exercises by dividing by two one-digit divisors.

Elementary Exercises

1. 2,072 ÷ 28

$$\begin{array}{r} 7\ 4 \\ 7\overline{)\,5\ 1\,{}^{2}8} \\ 4\overline{)\,2,0\ 7\,{}^{3}2} \end{array}$$

2. 2,988 ÷ 36

3. 2,187 ÷ 81

4. 1,560 ÷ 24

5. 4,272 ÷ 48

6. 2,272 ÷ 32

7. 1,764 ÷ 42

8. 1,848 ÷ 56

9. 1,197 ÷ 63

10. 3,672 ÷ 54

11. 1,325 ÷ 25

12. 1,746 ÷ 18

13. 1,421 ÷ 49

14. 2,176 ÷ 64

Brain Builders

1. 41,944 ÷ 56

2. 12,996 ÷ 36

3. 57,960 ÷ 72

4. 8,910 ÷ 45

5. 15,632 ÷ 16

6. 9,135 ÷ 21

7. 18,410 ÷ 35

8. 11,928 ÷ 42

(See solutions on page 203.)

Trick 16: Place-Value Multiplication: Two-Digit by One-Digit

Strategy: In this trick, and the next one, we are going to focus on what each digit within a number represents. For example, in the number 49, the 4 represents four tens, and the 9 represents nine ones. So 49 could be expressed as $(4 \times 10) + (9 \times 1)$, or $40 + 9$. As you'll see below, place-value multiplication is best performed from left to right. As such, it is easily accomplished (with practice) in one's head.

Elementary Example 1

13 × 7

Step 1. Recast 13, emphasizing place value: $10 + 3$.

Step 2. Multiply 7 by 10: $7 \times 10 = 70$.

Step 3. Multiply 7 by 3: $7 \times 3 = 21$.

Step 4. Add the 70 and the 21: $70 + 21 = 91$ (answer).

Thought Process Summary

$$\begin{array}{r}13\\ \times\ 7\\ \hline\end{array} \rightarrow \begin{array}{r}(10+3)\\ \times\qquad 7\\ \hline\end{array} \rightarrow \begin{array}{r}10\\ \times\ 7\\ \hline 70\end{array} + \begin{array}{r}3\\ \times 7\\ \hline 21\end{array} \rightarrow \begin{array}{r}70\\ +\ 21\\ \hline 91\end{array}$$

Elementary Example 2

26 × 5

Step 1. Recast 26, emphasizing place value: $20 + 6$.

Step 2. Multiply 5 by 20: $5 \times 20 = 100$.

Step 3. Multiply 5 by 6: $5 \times 6 = 30$.

Step 4. Add the 100 and the 30: $100 + 30 = 130$ (answer).

Thought Process Summary

$$\begin{array}{r} 26 \\ \times\ 5 \\ \hline \end{array} \rightarrow \begin{array}{r} (20 + 6) \\ \times \qquad 5 \\ \hline \end{array} \rightarrow \begin{array}{r} 20 \\ \times\ 5 \\ \hline 100 \end{array} + \begin{array}{r} 6 \\ \times\ 5 \\ \hline 30 \end{array} \rightarrow \begin{array}{r} 100 \\ +\ 30 \\ \hline 130 \end{array}$$

Brain Builder

68 × 7

Step 1. Recast 68, emphasizing place value: 60 + 8.

Step 2. Multiply 7 by 60: 7 × 60 = 420.

Step 3. Multiply 7 by 8: 7 × 8 = 56.

Step 4. Add the 420 and the 56: 420 + 56 = 476 (answer).

Thought Process Summary

$$\begin{array}{r} 68 \\ \times\ 7 \\ \hline \end{array} \rightarrow \begin{array}{r} (60 + 8) \\ \times \qquad 7 \\ \hline \end{array} \rightarrow \begin{array}{r} 60 \\ \times\ 7 \\ \hline 420 \end{array} + \begin{array}{r} 8 \\ \times\ 7 \\ \hline 56 \end{array} \rightarrow \begin{array}{r} 420 \\ +\ 56 \\ \hline 476 \end{array}$$

Food for Thought: The key to mastering this trick is being able to "store" the product of the first multiplication while performing the second one. Although it may feel awkward at first, place-value multiplication is really a superb way to perform mental math. Remember, practice makes perfect!

Practice Problems

Multiply the tens digit first when working these exercises.

Elementary Exercises

1. $\begin{array}{r} 24 \\ \times\ 8 \\ \hline \end{array}$ **192 (160 + 32)**

2. $\begin{array}{r} 43 \\ \times\ 5 \\ \hline \end{array}$

3. $\begin{array}{r} 15 \\ \times\ 7 \\ \hline \end{array}$

4. $\begin{array}{r} 37 \\ \times\ 3 \\ \hline \end{array}$

5. $\begin{array}{r} 27 \\ \times\ 6 \\ \hline \end{array}$

6. $\begin{array}{r} 49 \\ \times\ 4 \\ \hline \end{array}$

7. $\begin{array}{r} 18 \\ \times\ 9 \\ \hline \end{array}$

8. $\begin{array}{r} 34 \\ \times\ 7 \\ \hline \end{array}$

9. $\begin{array}{r} 45 \\ \times\ 8 \\ \hline \end{array}$

11. $\begin{array}{r} 48 \\ \times\ 3 \\ \hline \end{array}$

13. $\begin{array}{r} 55 \\ \times\ 4 \\ \hline \end{array}$

10. $\begin{array}{r} 38 \\ \times\ 6 \\ \hline \end{array}$

12. $\begin{array}{r} 34 \\ \times\ 5 \\ \hline \end{array}$

14. $\begin{array}{r} 19 \\ \times\ 9 \\ \hline \end{array}$

Brain Builders

1. $\begin{array}{r} 87 \\ \times\ 5 \\ \hline \end{array}$

3. $\begin{array}{r} 57 \\ \times\ 7 \\ \hline \end{array}$

5. $\begin{array}{r} 72 \\ \times\ 9 \\ \hline \end{array}$

7. $\begin{array}{r} 67 \\ \times\ 5 \\ \hline \end{array}$

2. $\begin{array}{r} 63 \\ \times\ 8 \\ \hline \end{array}$

4. $\begin{array}{r} 95 \\ \times\ 4 \\ \hline \end{array}$

6. $\begin{array}{r} 59 \\ \times\ 6 \\ \hline \end{array}$

8. $\begin{array}{r} 94 \\ \times\ 8 \\ \hline \end{array}$

(See solutions on page 204.)

Trick 17: Place-Value Multiplication: Three-Digit by One-Digit

Strategy: Make sure you fully understand Trick 16 before proceeding with this one. The technique is the same, except that we will introduce a hundreds digit into the calculation. Again, we will focus on what each digit within a number represents. So a number such as 835 could be expressed as (8 × 100) + (3 × 10) + (5 × 1) or 800 + 30 + 5. Let's take a look at this expanded version of multiplying from left to right.

Elementary Example 1

137 × 5

Step 1. Recast 137, emphasizing place value: 100 + 30 + 7.

Step 2. Multiply 5 by 100: 5 × 100 = 500.

Step 3. Multiply 5 by 30: 5 × 30 = 150.

Step 4. Multiply 5 by 7: 5 × 7 = 35.

Step 5. Add the 500, 150, and 35: 500 + 150 + 35 = 685 (answer).

Thought Process Summary

$$\begin{array}{r} 137 \\ \times\ 5 \\ \hline \end{array} \rightarrow \begin{array}{l} (100 + 30 + 7) \\ \times \qquad\qquad\quad 5 \\ \hline \end{array} \rightarrow \begin{array}{r} 100 \\ \times\ 5 \\ \hline 500 \end{array} + \begin{array}{r} 30 \\ \times\ 5 \\ \hline 150 \end{array} + \begin{array}{r} 7 \\ \times 5 \\ \hline 35 \end{array} \rightarrow \begin{array}{r} 500 \\ 150 \\ +\ 35 \\ \hline 685 \end{array}$$

Elementary Example 2

428 × 3

Step 1. Recast 428, emphasizing place value: 400 + 20 + 8.

Step 2. Multiply 3 by 400: 3 × 400 = 1,200.

Step 3. Multiply 3 by 20: 3 × 20 = 60.

Step 4. Multiply 3 by 8: 3 × 8 = 24.

Step 5. Add the 1,200, 60, and 24: 1,200 + 60 + 24 = 1,284 (answer).

Thought Process Summary

$$\begin{array}{r} 428 \\ \times\ 3 \\ \hline \end{array} \rightarrow \begin{array}{l} (400 + 20 + 8) \\ \times \qquad\qquad\quad 3 \\ \hline \end{array} \rightarrow \begin{array}{r} 400 \\ \times\ 3 \\ \hline 1{,}200 \end{array} + \begin{array}{r} 20 \\ \times\ 3 \\ \hline 60 \end{array} + \begin{array}{r} 8 \\ \times 3 \\ \hline 24 \end{array} \rightarrow \begin{array}{r} 1{,}200 \\ 60 \\ +\ 24 \\ \hline 1{,}284 \end{array}$$

Brain Builder

736 × 8

Step 1. Recast 736, emphasizing place value: 700 + 30 + 6.

Step 2. Multiply 8 by 700: 8 × 700 = 5,600.

Step 3. Multiply 8 by 30: 8 × 30 = 240.

Step 4. Multiply 8 by 6: 8 × 6 = 48.

Step 5. Add the 5,600, 240, and 48: 5,600 + 240 + 48 = 5,888 (answer).

Thought Process Summary

$$\begin{array}{r} 736 \\ \times\ 8 \\ \hline \end{array} \rightarrow \begin{array}{l} (700 + 30 + 6) \\ \times \qquad\qquad\quad 8 \\ \hline \end{array} \rightarrow \begin{array}{r} 700 \\ \times\ 8 \\ \hline 5{,}600 \end{array} + \begin{array}{r} 30 \\ \times\ 8 \\ \hline 240 \end{array} + \begin{array}{r} 6 \\ \times 8 \\ \hline 48 \end{array} \rightarrow \begin{array}{r} 5{,}600 \\ 240 \\ +\ 48 \\ \hline 5{,}888 \end{array}$$

Food for Thought: Notice that we are working from left to right, starting with the "big numbers" and ending with the small ones. The reason we are doing this is that it is much easier to begin with large numbers and to then add smaller ones. Using our Brain Builder example above, try starting with 48 and then adding 240 and finally 5,600. You'll see that it is far more difficult to begin with small numbers and then add larger ones.

Practice Problems

Multiply from left to right to solve these problems.

Elementary Exercises

1. 123×4 = **492** **(400 + 80 + 12)**
2. 429×5
3. 226×3
4. 315×6
5. 143×7
6. 257×4
7. 124×8
8. 471×5
9. 185×9
10. 462×3
11. 277×6
12. 478×2
13. 237×7
14. 495×5

Brain Builders

1. 812×7
2. 692×5
3. 583×6
4. 922×4
5. 731×9
6. 694×3
7. 557×8
8. 748×5

(See solutions on page 204.)

Trick 18: Dividing by Regrouping I

Strategy: Very often a calculation is best accomplished in two or more steps rather than one. Such is the case with many of the tricks in this book (including this one). Trick 18 is best applied when the number being divided is a bit over 100, or a multiple of 100. A calculation such as 115 ÷ 5 would qualify. First, split the number being divided into two so that one part is 100. Then divide both parts of the number by the divisor and add the results. Let's look at some illustrations of this trick below.

Elementary Example 1

118 ÷ 2

Step 1. Split 118 into 100 and 18: 118 = 100 + 18.

Step 2. Divide 100 by 2: 100 ÷ 2 = 50.

Step 3. Divide 18 by 2: 18 ÷ 2 = 9.

Step 4. Add the 50 and the 9: 50 + 9 = 59 (answer).

Thought Process Summary

$$118 \div 2 \rightarrow (100 + 18) \div 2 \rightarrow \begin{array}{r} (100 \div 2) = 50 \\ + (18 \div 2) = \underline{\ \ 9} \\ 59 \end{array}$$

Elementary Example 2

112 ÷ 4

Step 1. Split 112 into 100 and 12: 112 = 100 + 12.

Step 2. Divide 100 by 4: 100 ÷ 4 = 25.

Step 3. Divide 12 by 4: 12 ÷ 4 = 3.

Step 4. Add the 25 and the 3: 25 + 3 = 28 (answer).

Thought Process Summary

$$112 \div 4 \rightarrow (100 + 12) \div 4 \rightarrow \begin{array}{r} (100 \div 4) = 25 \\ + (12 \div 4) = \underline{\ 3} \\ 28 \end{array}$$

Brain Builder

345 ÷ 15

Step 1. Split 345 into 300 and 45: 345 = 300 + 45.

Step 2. Divide 300 by 15: 300 ÷ 15 = 20.

Step 3. Divide 45 by 15: 45 ÷ 15 = 3.

Step 4. Add the 20 and the 3: 20 + 3 = 23 (answer).

Thought Process Summary

$$345 \div 15 \rightarrow (300 + 45) \div 15 \rightarrow \begin{array}{r} (300 \div 15) = 20 \\ + (45 \div 15) = \underline{\ 3} \\ 23 \end{array}$$

Food for Thought: This trick would work when dividing, let's say, 165 by 5, even though 165 is more than just a little over 100. It would just take a bit longer to solve than the examples above. In addition, using this trick will not guarantee that you will obtain a whole-number answer (no trick will). Nevertheless, it is a good, solid trick. Try it for the calculation 114 ÷ 4. Even though the solution is not a whole number, the trick still works like a charm.

Practice Problems

The secret is to split the dividend, divide twice, then add.

Elementary Exercises

1. 115 ÷ 5 =
   (100 ÷ 5) + (15 ÷ 5) = 23
2. 112 ÷ 2 =
3. 116 ÷ 4 =
4. 125 ÷ 5 =
5. 114 ÷ 2 =
6. 132 ÷ 4 =

7. $210 \div 5 =$

8. $128 \div 4 =$

9. $116 \div 2 =$

10. $212 \div 4 =$

11. $315 \div 5 =$

12. $214 \div 2 =$

13. $608 \div 4 =$

14. $435 \div 5 =$

Brain Builders

1. $575 \div 25 =$

2. $675 \div 15 =$

3. $770 \div 35 =$

4. $390 \div 15 =$

5. $945 \div 45 =$

6. $600 \div 25 =$

7. $840 \div 35 =$

8. $990 \div 15 =$

(See solutions on page 204.)

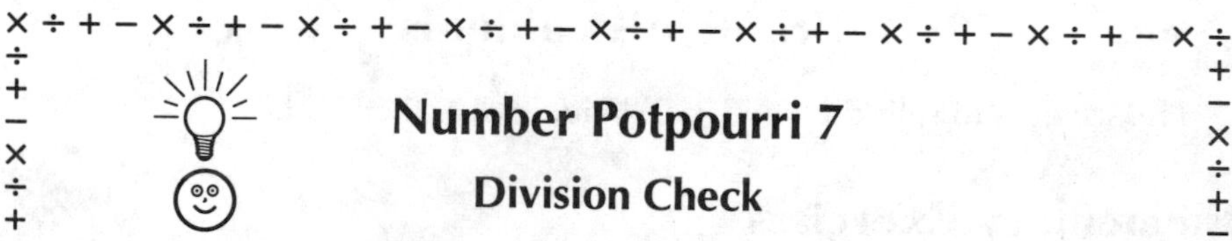

Number Potpourri 7

Division Check

When you've just divided one number by another, you can check your answer really quickly by working backward. Let's say you just divided 612 by 36 and came up with the answer 17. At this point, simply take the 17 and multiply it by 36 to see if you get 612. This method works especially well when you're dividing large numbers on your calculator.

Trick 19: Dividing by Regrouping II

Strategy: This trick is similar to Trick 18 in that the number to be divided is split into two parts. However, in these cases, the divisor will not divide evenly into 100, making the calculation a bit more challenging. Moreover, as you will see in our examples below, a little more imagination is required than was needed for Trick 18 to accomplish this regrouping technique.

Elementary Example 1

108 ÷ 3

Step 1. Determine a number near 108 that 3 divides comfortably and evenly into: 99 (our choice).

Step 2. Split 108 into 99 and 9: 108 = 99 + 9.

Step 3. Divide 99 by 3: 99 ÷ 3 = 33.

Step 4. Divide 9 by 3: 9 ÷ 3 = 3.

Step 5. Add the 33 and the 3: 33 + 3 = 36 (answer).

Thought Process Summary

$$108 \div 3 \rightarrow (99 + 9) \div 3 \rightarrow \begin{array}{r} (99 \div 3) = 33 \\ + (9 \div 3) = \underline{3} \\ 36 \end{array}$$

Elementary Example 2

154 ÷ 7

Step 1. Determine a number near 154 that 7 divides comfortably and evenly into: 140 (our choice).

Step 2. Split 154 into 140 and 14: 154 = 140 + 14.

Step 3. Divide 140 by 7: 140 ÷ 7 = 20.

Step 4. Divide 14 by 7: 14 ÷ 7 = 2.

Step 5. Add the 20 and the 2: 20 + 2 = 22 (answer).

Thought Process Summary

$$154 \div 7 \rightarrow (140 + 14) \div 7 \rightarrow \begin{array}{r} (140 \div 7) = 20 \\ + (14 \div 7) = \underline{2} \\ 22 \end{array}$$

Brain Builder

$403 \div 13$

Step 1. Determine a number near 403 that 13 divides comfortably and evenly into: 390 (our choice).

Step 2. Split 403 into 390 and 13: $403 = 390 + 13$.

Step 3. Divide 390 by 13: $390 \div 13 = 30$.

Step 4. Divide 13 by 13: $13 \div 13 = 1$.

Step 5. Add the 30 and the 1: $30 + 1 = 31$ (answer).

Thought Process Summary

$$403 \div 13 \rightarrow (390 + 13) \div 13 \rightarrow \begin{array}{r} (390 \div 13) = 30 \\ + (13 \div 13) = \underline{\ \ 1} \\ 31 \end{array}$$

Food for Thought: The examples presented in Tricks 18 and 19 are solved by first "falling short" and then "picking up the rest" to complete the calculation. In Tricks 24 and 25, however, we will "overshoot" to begin the division and then "backtrack" to complete it.

Practice Problems

You'll need to use a bit of imagination when splitting these dividends.

Elementary Exercises

1. $105 \div 3 =$
   $(99 \div 3) + (6 \div 3) = 35$
2. $138 \div 6 =$
3. $168 \div 7 =$
4. $144 \div 9 =$
5. $198 \div 6 =$
6. $156 \div 3 =$
7. $198 \div 9 =$
8. $245 \div 7 =$
9. $138 \div 3 =$

10. $378 \div 9 =$

11. $342 \div 6 =$

12. $301 \div 7 =$

13. $639 \div 9 =$

14. $192 \div 3 =$

Brain Builders

1. $252 \div 12 =$
2. $704 \div 22 =$
3. $182 \div 13 =$
4. $693 \div 11 =$
5. $588 \div 14 =$
6. $720 \div 16 =$
7. $765 \div 15 =$
8. $768 \div 24 =$

(See solutions on page 205.)

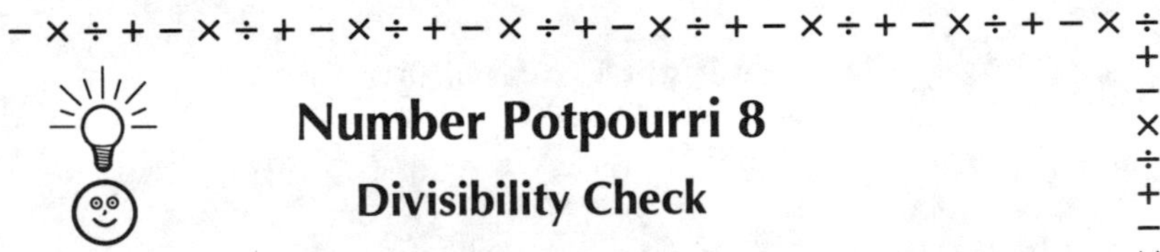

Number Potpourri 8

Divisibility Check

A number is evenly divisible by 4 if the last two digits of the number are evenly divisible by 4, or if the number ends in 00. For example, 7,256 is evenly divisible by 4 because 4 divides evenly into 56. A number is evenly divisible by 8 if the last three digits of the number are evenly divisible by 8, or if the number ends in 000. For example, 513,216 is evenly divisible by 8 because 8 divides evenly into 216.

Trick 20: Multiplying by 21, 31, 41, etc.

Strategy: Multiplication problems with 21, 31, and the like lend themselves very well to place-value multiplication. To multiply a number by 21, for example, you would first multiply the number by 20 and then add the number itself to produce the answer. Though accomplished essentially in two steps, multiplying by a number ending in 1 is far simpler to accomplish if converted (at least, in part) to multiplying by a number ending in zero.

Elementary Example 1

35 × 21

Step 1. Multiply 35 by 20: $35 \times 20 = 700$.

Step 2. Add 35 to the 700: $700 + 35 = 735$ (answer).

Thought Process Summary

$$\begin{array}{r} 35 \\ \times\ 21 \\ \hline \end{array} \rightarrow \begin{array}{r} 35 \\ \times\ 20 \\ \hline 700 \end{array} \rightarrow \begin{array}{r} 700 \\ +\ 35 \\ \hline 735 \end{array}$$

Elementary Example 2

16 × 31

Step 1. Multiply 16 by 30: $16 \times 30 = 480$.

Step 2. Add 16 to the 480: $480 + 16 = 496$ (answer).

Thought Process Summary

$$\begin{array}{r} 16 \\ \times\ 31 \\ \hline \end{array} \rightarrow \begin{array}{r} 16 \\ \times\ 30 \\ \hline 480 \end{array} \rightarrow \begin{array}{r} 480 \\ +\ 16 \\ \hline 496 \end{array}$$

Brain Builder

54 × 81

Step 1. Multiply 54 by 80: 54 × 80 = 4,320.

Step 2. Add 54 to the 4,320: 4,320 + 54 = 4,374 (answer).

Thought Process Summary

$$\begin{array}{r} 54 \\ \times\ 81 \\ \hline \end{array} \rightarrow \begin{array}{r} 54 \\ \times\ 80 \\ \hline 4{,}320 \end{array} \rightarrow \begin{array}{r} 4{,}320 \\ +\ \ \ 54 \\ \hline 4{,}374 \end{array}$$

Food for Thought: In effect, this trick takes a calculation such as 35 × 21 and converts it into 35 × (20 + 1), which in turn equals (35 × 20) + (35 × 1). In addition, Trick 20 will work when multiplying by numbers such as 2.1, 310, 0.41, and so forth. Simply multiply as illustrated above (ignoring the decimal point or right-hand zeros), and then insert a decimal point or tack on one or more zeros where appropriate. These examples were limited to basic whole numbers to simplify the explanation.

Practice Problems

Make sure to split the multiplier into a multiple of 10, plus 1.

Elementary Exercises

1. 45 × 21 =
   (45 × 20) + (45 × 1) = 945
2. 14 × 51 =
3. 24 × 31 =
4. 12 × 71 =
5. 25 × 41 =
6. 16 × 61 =

7. $15 \times 91 =$

8. $30 \times 81 =$

9. $13 \times 31 =$

10. $22 \times 51 =$

11. $44 \times 21 =$

12. $32 \times 61 =$

13. $17 \times 41 =$

14. $35 \times 81 =$

Brain Builders

1. $85 \times 21 =$

2. $63 \times 71 =$

3. $55 \times 41 =$

4. $52 \times 61 =$

5. $93 \times 31 =$

6. $65 \times 81 =$

7. $84 \times 51 =$

8. $72 \times 91 =$

(See solutions on page 205.)

Number Potpourri 9

Multiplication Check

When you've just multiplied two numbers, you can check your answer really quickly by working backward. Let's say you just multiplied 37 by 21 and came up with the answer 777. At this point, simply take the 777 and divide by 21 to see if you get 37. Or you could divide the 777 by 37 to see if you get 21. This method works especially well when you're multiplying large numbers on your calculator.

Trick 21: Multiplying by 19, 29, 39, etc.

Strategy: This trick is similar to the last one in that we will multiply by a multiple of 10 to simplify the procedure. To multiply a number by 19, for example, first multiply the number by 20 (the next multiple of 10), and then deduct the number itself to produce the answer. In effect, you are multiplying by (20 − 1). Let's look at some examples below.

Elementary Example 1

8 × 19

Step 1. Multiply 8 by 20: 8 × 20 = 160.

Step 2. Subtract 8 from the 160: 160 − 8 = 152 (answer).

Thought Process Summary

$$\begin{array}{r} 8 \\ \times\ 19 \\ \hline \end{array} \rightarrow \begin{array}{r} 8 \\ \times\ 20 \\ \hline 160 \end{array} \rightarrow \begin{array}{r} 160 \\ -\ \ 8 \\ \hline 152 \end{array}$$

Elementary Example 2

6 × 29

Step 1. Multiply 6 by 30: 6 × 30 = 180.

Step 2. Subtract 6 from the 180: 180 − 6 = 174 (answer).

Thought Process Summary

$$\begin{array}{r} 6 \\ \times\ 29 \\ \hline \end{array} \rightarrow \begin{array}{r} 6 \\ \times\ 30 \\ \hline 180 \end{array} \rightarrow \begin{array}{r} 180 \\ -\ \ 6 \\ \hline 174 \end{array}$$

Brain Builder

74 × 49

Step 1. Multiply 74 by 50: 74 × 50 = 3,700.

Step 2. Subtract 74 from the 3,700: 3,700 − 74 = 3,626 (answer).

Thought Process Summary

$$\begin{array}{r} 74 \\ \underline{\times\ 49} \end{array} \rightarrow \begin{array}{r} 74 \\ \underline{\times\ 50} \\ 3{,}700 \end{array} \rightarrow \begin{array}{r} 3{,}700 \\ \underline{-\quad 74} \\ 3{,}626 \end{array}$$

Food for Thought: Like Trick 20, this trick will work when multiplying by numbers such as 1.9, 290, 0.39, and so forth. Simply multiply as illustrated above (ignoring the decimal point or right-hand zeros), and then insert a decimal point or tack on one or more zeros where appropriate. Our examples were limited to basic whole numbers to simplify the explanation.

Practice Problems

Round up the multiplier as the first step to solving these problems.

Elementary Exercises

1. 7 × 19 =
   (7 × 20) − (7 × 1) = 133
2. 8 × 49 =
3. 6 × 29 =
4. 5 × 79 =
5. 9 × 59 =
6. 4 × 99 =
7. 7 × 39 =
8. 3 × 89 =
9. 5 × 69 =
10. 8 × 49 =
11. 9 × 19 =
12. 6 × 59 =
13. 7 × 29 =
14. 4 × 39 =

Brain Builders

1. $12 \times 49 =$
2. $30 \times 89 =$
3. $11 \times 69 =$
4. $20 \times 79 =$
5. $15 \times 39 =$
6. $24 \times 19 =$
7. $40 \times 59 =$
8. $13 \times 29 =$

(See solutions on page 205.)

Number Challenge 1

1. Approximately what is the hottest naturally occurring air temperature (in the shade) ever recorded?

   a. 119°F b. 136°F c. 152°F d. 177°F

2. Approximately what is the coldest naturally occurring air temperature ever recorded on earth?

   a. −128°F b. −213°F c. −387°F d. −459°F

(See answers on page 222.)

Trick 22: Multiplying by 12

Strategy: The number 12 is commonly used in calculations, in part because things are often measured in dozens. Unfortunately, most multiplications involving 12, such as 15×12, are slightly out of reach for most people. Fortunately, the number 12 can be expressed (factored) in a number of different ways: 3×4, $2 \times 2 \times 3$, 6×2, $3 \times 2 \times 2$, and so forth. Therefore, our 15×12 problem could be restated more simply as $15 \times 4 \times 3$, which equals 60×3, or 180. Or it could be restated as $15 \times 2 \times 6$, which equals 30×6, or 180. Exactly how you restate the number 12 is a matter of practice and creativity. Let's take a look at some more examples.

Elementary Example 1

14 × 12

Step 1. Recast the computation as $14 \times 6 \times 2$ (or other factors of 12).

Step 2. Multiply the first two numbers, then the third.

Thought Process Summary

$14 \times 12 = 14 \times 6 \times 2 = 84 \times 2 = 168$ (answer)

Elementary Example 2

21 × 12

Step 1. Recast the computation as $21 \times 3 \times 2 \times 2$.

Step 2. Multiply the numbers in order, from left to right.

Thought Process Summary

$21 \times 12 = 21 \times 3 \times 2 \times 2 = 63 \times 2 \times 2 = 126 \times 2 = 252$ (answer)

Brain Builder

65 × 12

Step 1. Recast the computation as $65 \times 4 \times 3$.

Step 2. Multiply the first two numbers, then the third.

Thought Process Summary

$65 \times 12 = 65 \times 4 \times 3 = 260 \times 3 = 780$ (answer)

Food for Thought: Often there is more than one way to perform a calculation rapidly. For example, you could multiply a number by 12 by multiplying it by 10, then multiplying it by 2, and then adding the results. Trick 22 simply represents a different way to arrive at the same result.

Practice Problems

Split 12 into factors to simplify these multiplication problems.

Elementary Exercises

1. $15 \times 12 =$ **$(15 \times 4) \times 3 = 180$**
2. $9 \times 12 =$
3. $13 \times 12 =$
4. $17 \times 12 =$
5. $22 \times 12 =$
6. $16 \times 12 =$
7. $23 \times 12 =$
8. $19 \times 12 =$
9. $26 \times 12 =$
10. $18 \times 12 =$
11. $24 \times 12 =$
12. $14 \times 12 =$
13. $21 \times 12 =$
14. $25 \times 12 =$

Brain Builders

1. $35 \times 12 =$
2. $32 \times 12 =$
3. $75 \times 12 =$
4. $52 \times 12 =$
5. $45 \times 12 =$
6. $27 \times 12 =$
7. $55 \times 12 =$
8. $36 \times 12 =$

(See solutions on page 206.)

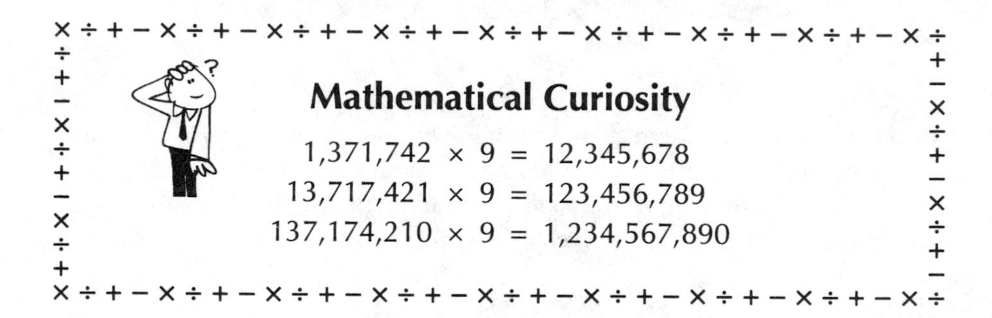

Trick 23: Dividing by 12

Strategy: We are going to divide by 12 in exactly the same way we multiplied by 12 in Trick 22—by expressing (factoring) the number 12 as $2 \times 2 \times 3$, 4×3, 6×2, and so forth. Performing division by 12, however, is a bit more difficult because we are basically working backward. For example, we could restate $96 \div 12$ as $96 \div 2 \div 6$, which equals $48 \div 6$, or 8. Read on for more illustrations of this useful trick.

Elementary Example 1

168 ÷ 12

Step 1. Recast the computation as $168 \div 2 \div 2 \div 3$ (our choice).

Step 2. Divide, in order, from left to right.

Thought Process Summary

$168 \div 12 = 168 \div 2 \div 2 \div 3 = 84 \div 2 \div 3 = 42 \div 3 = 14$ (answer)

Elementary Example 2

108 ÷ 12

Step 1. Recast the computation as 108 ÷ 2 ÷ 6.

Step 2. Divide, in order, from left to right.

Thought Process Summary

108 ÷ 12 = 108 ÷ 2 ÷ 6 = 54 ÷ 6 = 9 (answer)

Brain Builder

312 ÷ 12

Step 1. Recast the computation as 312 ÷ 3 ÷ 2 ÷ 2.

Step 2. Divide, in order, from left to right.

Thought Process Summary

312 ÷ 12 = 312 ÷ 3 ÷ 2 ÷ 2 = 104 ÷ 2 ÷ 2 = 52 ÷ 2 = 26 (answer)

Food for Thought: Very often you can use two rapid math techniques at once to perform a computation. For example, did you use Trick 18 (Dividing by Regrouping I) to divide 312 by 3 in our Brain Builder example? That is, you would first divide 300 by 3, then divide 12 by 3, and add the results. Don't hesitate to use any previously learned trick to help solve the problem at hand!

Practice Problems

Split 12 into factors to simplify these division problems.

Elementary Exercises

1. 156 ÷ 12 =
   156 ÷ 2 ÷ 2 ÷ 3 = 13
2. 204 ÷ 12 =
3. 168 ÷ 12 =
4. 108 ÷ 12 =
5. 276 ÷ 12 =
6. 192 ÷ 12 =

7. $288 \div 12 =$

8. $300 \div 12 =$

9. $180 \div 12 =$

10. $252 \div 12 =$

11. $216 \div 12 =$

12. $96 \div 12 =$

13. $228 \div 12 =$

14. $264 \div 12 =$

Brain Builders

1. $540 \div 12 =$

2. $324 \div 12 =$

3. $900 \div 12 =$

4. $432 \div 12 =$

5. $384 \div 12 =$

6. $420 \div 12 =$

7. $312 \div 12 =$

8. $780 \div 12 =$

(See solutions on page 206.)

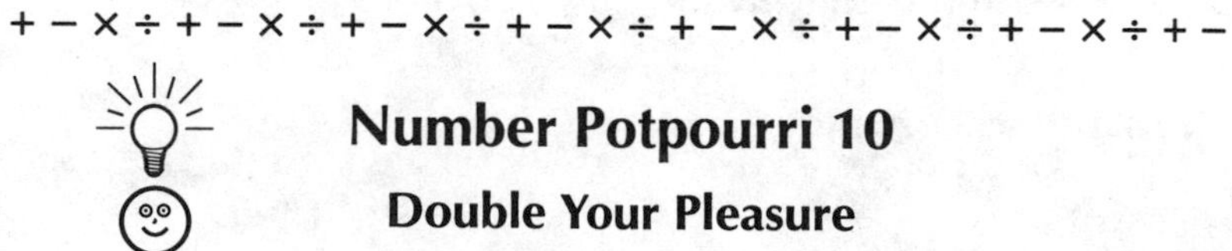

Number Potpourri 10

Double Your Pleasure

Take any three-digit number and repeat it to form a six-digit number. So a number such as 726 would become 726,726. Then divide the six-digit number by (in any order) 7, 11, and 13. You'll arrive at the original three-digit number every time (e.g., $726{,}726 \div 7 \div 11 \div 13 = 726$). Try it on another three-digit number, and see for yourself!

Trick 24: Dividing by Augmenting I

Strategy: This trick works best when the number being divided is a little under 100 (or a multiple of 100). Think of the number being divided as 100 minus something. So 96, for example, becomes 100 − 4. Perform division on the two parts — the 100 and the 4 — and then subtract. Let's take a closer look at this effective trick.

Elementary Example 1

96 ÷ 4

Step 1. Restate 96 as 100 minus 4: $96 = 100 - 4$.

Step 2. Divide 100 by 4: $100 \div 4 = 25$.

Step 3. Divide 4 by 4: $4 \div 4 = 1$.

Step 4. Subtract the 1 from the 25: $25 - 1 = 24$ (answer).

Thought Process Summary

$$96 \div 4 \rightarrow (100 - 4) \div 4 \rightarrow \begin{array}{r} (100 \div 4) = 25 \\ -\ (4 \div 4) = \underline{\ \ 1} \\ 24 \end{array}$$

Elementary Example 2

190 ÷ 5

Step 1. Restate 190 as 200 minus 10: $190 = 200 - 10$.

Step 2. Divide 200 by 5: $200 \div 5 = 40$.

Step 3. Divide 10 by 5: $10 \div 5 = 2$.

Step 4. Subtract the 2 from the 40: $40 - 2 = 38$ (answer).

Thought Process Summary

$$190 \div 5 \rightarrow (200 - 10) \div 5 \rightarrow \begin{array}{r} (200 \div 5) = 40 \\ -\ (10 \div 5) = \underline{\ 2} \\ 38 \end{array}$$

Brain Builder

285 ÷ 15

Step 1. Restate 285 as 300 minus 15: 285 = 300 − 15.
Step 2. Divide 300 by 15: 300 ÷ 15 = 20.
Step 3. Divide 15 by 15: 15 ÷ 15 = 1.
Step 4. Subtract the 1 from the 20: 20 − 1 = 19 (answer).

Thought Process Summary

$$285 \div 15 \rightarrow (300 - 15) \div 15 \rightarrow \begin{array}{r} (300 \div 15) = 20 \\ -\ (15 \div 15) = \underline{\ 1} \\ 19 \end{array}$$

Food for Thought: This trick would work well when dividing, let's say, 76 by 4, even though 76 is well under 100. It would just take a bit longer to solve than the examples above.

Practice Problems

Begin these exercises by rounding up to the nearest multiple of 100.

Elementary Exercises

1. 90 ÷ 5 =
   (100 ÷ 5) − (10 ÷ 5) = 18
2. 98 ÷ 2 =
3. 92 ÷ 4 =
4. 194 ÷ 2 =
5. 192 ÷ 4 =
6. 185 ÷ 5 =

7. 298 ÷ 2 =

8. 288 ÷ 4 =

9. 290 ÷ 5 =

10. 392 ÷ 2 =

11. 392 ÷ 4 =

12. 385 ÷ 5 =

13. 494 ÷ 2 =

14. 495 ÷ 5 =

Brain Builders

1. 475 ÷ 25 =

2. 810 ÷ 45 =

3. 570 ÷ 15 =

4. 665 ÷ 35 =

5. 405 ÷ 15 =

6. 925 ÷ 25 =

7. 630 ÷ 35 =

8. 885 ÷ 15 =

(See solutions on page 206.)

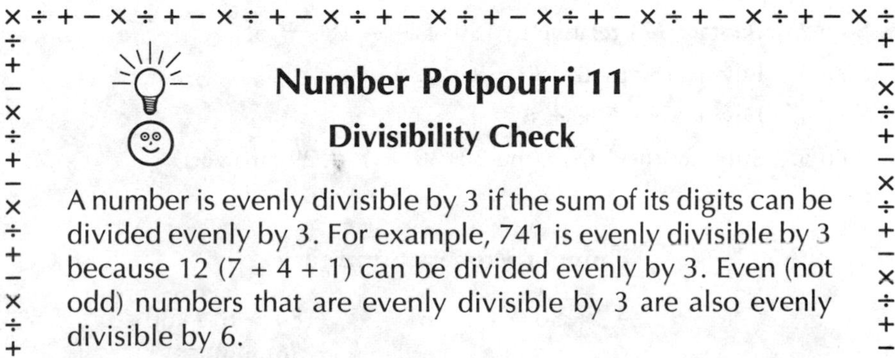

Number Potpourri 11

Divisibility Check

A number is evenly divisible by 3 if the sum of its digits can be divided evenly by 3. For example, 741 is evenly divisible by 3 because 12 (7 + 4 + 1) can be divided evenly by 3. Even (not odd) numbers that are evenly divisible by 3 are also evenly divisible by 6.

Trick 25: Dividing by Augmenting II

Strategy: This trick involves the same approach as Trick 24, except that we will now address some slightly more complicated cases. These cases will not be limited to amounts just under 100 (or a multiple of 100), and they require a bit more imagination than the previous trick. For example, we will view a calculation such as 141 ÷ 3 as much more "user-friendly": (150 − 9) ÷ 3. Let's look at some step-by-step examples below.

Elementary Example 1

141 ÷ 3

Step 1. Determine a number near 141 that 3 divides comfortably and evenly into: 150 (our choice).

Step 2. Restate 141 relative to 150: 141 = 150 − 9.

Step 3. Divide 150 by 3: 150 ÷ 3 = 50.

Step 4. Divide 9 by 3: 9 ÷ 3 = 3.

Step 5. Subtract the 3 from the 50: 50 − 3 = 47 (answer).

Thought Process Summary

$$141 \div 3 \rightarrow (150 - 9) \div 3 \rightarrow \begin{array}{r} (150 \div 3) = 50 \\ -\ (9 \div 3) = \underline{\ \ 3} \\ 47 \end{array}$$

Elementary Example 2

174 ÷ 6

Step 1. Determine a number near 174 that 6 divides comfortably and evenly into: 180.

Step 2. Restate 174 relative to 180: 174 = 180 − 6.

Step 3. Divide 180 by 6: 180 ÷ 6 = 30.

Step 4. Divide 6 by 6: 6 ÷ 6 = 1.

Step 5. Subtract the 1 from the 30: 30 − 1 = 29 (answer).

Thought Process Summary

$$174 \div 6 \rightarrow (180 - 6) \div 6 \rightarrow \begin{array}{r} (180 \div 6) = 30 \\ -\ (6 \div 6) = \underline{\ \ 1} \\ 29 \end{array}$$

Brain Builder

418 ÷ 11

Step 1. Determine a number near 418 that 11 divides comfortably and evenly into: 440.

Step 2. Restate 418 relative to 440: 418 = 440 − 22.

Step 3. Divide 440 by 11: 440 ÷ 11 = 40.

Step 4. Divide 22 by 11: 22 ÷ 11 = 2.

Step 5. Subtract the 2 from the 40: 40 − 2 = 38 (answer).

Thought Process Summary

$$418 \div 11 \rightarrow (440 - 22) \div 11 \rightarrow \begin{array}{r} (440 \div 11) = 40 \\ -\ (22 \div 11) = \underline{\ 2} \\ 38 \end{array}$$

Food for Thought: Tricks 24 and 25 involve divisions that are easiest to perform by overshooting the number you're dividing into and then backtracking. On the other hand, Tricks 18 and 19 approach division problems by undershooting the number you're dividing into and then making up the difference. For example, we could have approached our 141 ÷ 3 problem instead as (120 ÷ 3) + (21 ÷ 3), to produce the same answer, 47.

Practice Problems

Remember to overshoot and backtrack when solving these exercises.

Elementary Exercises

1. 144 ÷ 3 =
   (150 ÷ 3) − (6 ÷ 3) = 48
2. 133 ÷ 7 =
3. 174 ÷ 6 =
4. 162 ÷ 9 =
5. 117 ÷ 3 =
6. 174 ÷ 6 =
7. 432 ÷ 9 =
8. 291 ÷ 3 =

9. $406 \div 7 =$

10. $294 \div 6 =$

11. $171 \div 3 =$

12. $891 \div 9 =$

13. $483 \div 7 =$

14. $234 \div 3 =$

Brain Builders

1. $108 \div 12 =$

2. $528 \div 11 =$

3. $585 \div 15 =$

4. $1{,}386 \div 14 =$

5. $768 \div 16 =$

6. $247 \div 13 =$

7. $814 \div 22 =$

8. $1{,}649 \div 17 =$

(See solutions on page 207.)

Number Potpourri 12

Divide and Conquer

Here's a quick way to divide the numbers 1 through 10 by 11: Just multiply by 9, and write the answer as a repeating decimal. For example, to compute $3 \div 11$, just multiply the 3 by 9 (which equals 27), and write the answer as 0.272727 . . . . Similarly, $10 \div 11 = 0.909090 \ldots$, because $10 \times 9 = 90$. There is one calculation that deserves special attention: $1 \div 11$. Just multiply the 1 by 9 (which, of course, equals 9), but write the answer as 0.090909 . . . .

Trick 26: Subtracting by Oversubtracting

Strategy: This trick works best when you are subtracting a number that is just under a multiple of 100. In these instances, it's best to subtract the next higher multiple of 100 and then add back the amount that was purposely oversubtracted. As you'll see below, this technique is a lot easier to apply than it sounds.

Elementary Example 1

251 − 85

Step 1. Determine the multiple of 100 that 85 is closest to: 100.

Step 2. Subtract 100 from 251: $251 - 100 = 151$.

Step 3. Add back the amount that was oversubtracted in Step 2 (15, or $100 - 85$): $151 + 15 = 166$ (answer).

Thought Process Summary

$$\begin{array}{r} 251 \\ -\ 85 \\ \hline \end{array} \rightarrow \begin{array}{r} 251 \\ -\ 100 \\ \hline 151 \end{array} \rightarrow \begin{array}{r} 151 \\ +\ 15 \\ \hline 166 \end{array}$$

Elementary Example 2

423 − 194

Step 1. Determine the multiple of 100 that 194 is closest to: 200.

Step 2. Subtract 200 from 423: $423 - 200 = 223$.

Step 3. Add back the amount that was oversubtracted in Step 2 (6, or $200 - 194$): $223 + 6 = 229$ (answer).

Thought Process Summary

$$\begin{array}{r} 423 \\ -\ 194 \\ \hline \end{array} \rightarrow \begin{array}{r} 423 \\ -\ 200 \\ \hline 223 \end{array} \rightarrow \begin{array}{r} 223 \\ +\ \ 6 \\ \hline 229 \end{array}$$

Brain Builder

966 − 578

Step 1. Determine the multiple of 100 that 578 is closest to: 600.

Step 2. Subtract 600 from 966: 966 − 600 = 366.

Step 3. Add back the amount that was oversubtracted in Step 2 (22, or 600 − 578): 366 + 22 = 388 (answer).

Thought Process Summary

$$\begin{array}{r} 966 \\ -\ 578 \\ \hline \end{array} \rightarrow \begin{array}{r} 966 \\ -\ 600 \\ \hline 366 \end{array} \rightarrow \begin{array}{r} 366 \\ +\ 22 \\ \hline 388 \end{array}$$

Food for Thought: As you may have guessed, this trick will also work with multiples of 10, 1,000, and so forth. For example, you could compute the answer to 93 − 48 by taking 50 from 93, and then adding back the 2 that was oversubtracted.

Practice Problems

When working these exercises, remember to add back after oversubtracting.

Elementary Exercises

1. 141 − 88 =
   (141 − 100) + 12 = 53
2. 265 − 96 =
3. 253 − 95 =
4. 431 − 185 =
5. 375 − 191 =
6. 511 − 287 =

7. $434 - 296 =$

8. $550 - 183 =$

9. $333 - 289 =$

10. $246 - 93 =$

11. $373 - 197 =$

12. $424 - 298 =$

13. $526 - 89 =$

14. $266 - 95 =$

Brain Builders

1. $825 - 275 =$

2. $746 - 399 =$

3. $963 - 585 =$

4. $634 - 92 =$

5. $712 - 188 =$

6. $851 - 478 =$

7. $901 - 383 =$

8. $1{,}041 - 395 =$

(See solutions on page 207.)

Parlor Trick 2: The Amazing Age-Divining Trick

Like our "Mystifying Missing-Digit Trick" (Parlor Trick 1), there are several different ways to perform this trick. Here is one of the more impressive ways.

In this trick, you are going to guess someone's age. Therefore, your subject must be someone whose age you don't know. First, make sure you have a calculator handy. Then ask your subject to write down any whole number, such as 827 or 93,661. It doesn't matter what size the number is, as long as it will fit on your calculator. You are not allowed to see the number, but make sure your subject shows it to your audience (assuming you have one). You can either turn your back or wear a blindfold (for special effect!).

Next, have your subject add up the digits of the number chosen, either in his or her head or using the calculator. For example, if the number chosen was 6,805, the sum of the digits will be 19 (6 + 8 + 0 + 5). Have your subject subtract the sum of the digits from the number itself. So in our example, we would subtract 19 from 6,805, producing 6,786. You, of course, are not allowed to view any of the numbers or calculations.

At this point, ask your subject to add his or her age to the above result. If your subject is 42, for example, he or she would calculate 6,786 + 42 = 6,828.

Next, ask your subject to slowly read aloud, and in random order, all of the digits of the above result. So in our example, he or she might read the digits as 2, 8, 6, and 8 (or in any other order). What you must do is add those digits in your head. In our example, you would arrive at the sum 24 (2 + 8 + 6 + 8).

Finally, you take the above sum and either add or subtract 9, or a multiple of 9, until you arrive at the person's apparent age. In our example above, you would add 9 to the 24, to produce 33. You then ask yourself, "Does my subject appear to be 33 years old, or should I add another 9 years?" Perhaps your subject is a teenager, and you would subtract 9 from the 24, to produce 15 as your subject's age.

In short, you must choose a subject whose age you can estimate within 9 years. If in doubt, just remember that if you guess 9 years too old, the person is probably going to become rather embarrassed and insulted. So you might want to guess the younger age — if you're wrong, your subject will be so flattered he or she won't really care that you messed up the trick. Simply give a second guess at 9 years older.

What's especially impressive about this trick is that at no time do you know what number was chosen by your subject, nor do you know anything about his or her age. What your audience will not realize is that you've "stacked the deck" by having your subject subtract the sum of the digits from the number chosen.

Summary of Steps

1. Have your subject write down a whole number.
2. Have the subject add up the digits of the number.
3. Have the subject subtract the digit-sum from the number itself.
4. Have the subject add his or her age to the above result.
5. Have the subject read aloud, in random order, all digits of the result produced in Step 4. You will add the digits in your head.
6. You then add or subtract 9, 18, 27, and so on, until you arrive at what you think your subject's age is.

Week 2 Quick Quiz

Let's see how many tricks from Week 2 you can remember and apply by taking this brief test. There's no time limit, but try to work through these items as rapidly as possible. Before you begin, glance at the computations and try to identify the trick that you could use. In some instances, however, you will be asked to perform a calculation in a certain manner. When you flip ahead to the solutions, you will see which trick was intended.

Elementary Exercises

1. $7\overline{)3,612}$

2. Divide by two one-digit divisors:
   $1{,}638 \div 42 =$

3. Multiply the tens digit first:
   $$\begin{array}{r} 35 \\ \times\ 8 \\ \hline \end{array}$$

4. Multiply the hundreds digit first:
   $$\begin{array}{r} 158 \\ \times\ \ 6 \\ \hline \end{array}$$

5. $210 \div 5 =$

6. $162 \div 3 =$

7. $15 \times 41 =$

8. $8 \times 39 =$

9. $13 \times 12 =$

10. $216 \div 12 =$

11. $392 \div 4 =$

12. $171 \div 3 =$

13. $273 - 95 =$

Brain Builders

1. $13\overline{)3,679}$

2. Multiply the tens digit first:

$$\begin{array}{r} 74 \\ \times\ 7 \\ \hline \end{array}$$

3. $800 \div 25 =$

4. $35 \times 81 =$

5. $45 \times 12 =$

6. $665 \div 35 =$

7. $612 - 188 =$

(See solutions on page 218.)

Number Challenge 2

1. The world's tallest building, the Sears Tower in Chicago, contains how many stories?

   a. 90 b. 102 c. 110 d. 125

2. The deepest spot in the world's oceans, the Mariana Trench, descends approximately how far?

   a. 850 feet c. 3.5 miles
   b. 1.25 miles d. 6.75 miles

(See answers on page 222.)

Week 3 More Advanced Rapid Math Tricks

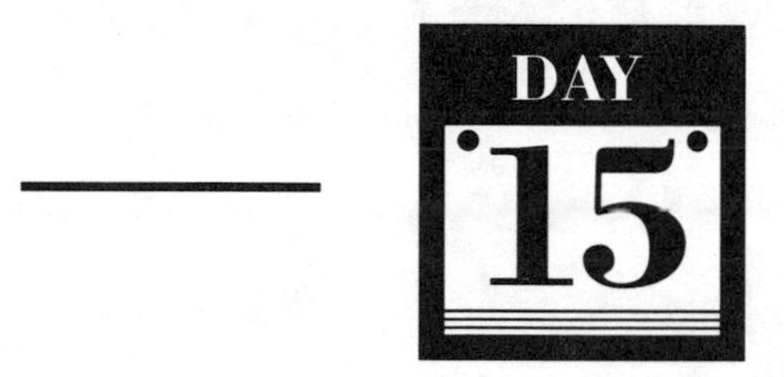

Trick 27: Multiplying with Numbers 11 Through 19

Strategy: This trick will work for calculations such as 14×19 and 16×11. The secret is to add either ones digit to the other number, multiply by 10, and then add the product of the ones digits. It's not really as complicated as it may seem, as you'll see in our examples below.

Elementary Example 1

13 × 16

Step 1. Add 6 to the 13: $13 + 6 = 19$.

Step 2. Multiply the 19 by 10: $19 \times 10 = 190$.

Step 3. Multiply together the ones digits: $3 \times 6 = 18$.

Step 4. Add the 18 to the 190: $190 + 18 = 208$ (answer).

Thought Process Summary

$$\begin{array}{r} 13 \\ \times\ 16 \\ \hline \end{array} \rightarrow \begin{array}{r} 13 \\ +\ 6 \\ \hline 19 \end{array} \rightarrow \begin{array}{r} 19 \\ \times\ 10 \\ \hline 190 \end{array} \rightarrow \begin{array}{r} 3 \\ \times\ 6 \\ \hline 18 \end{array} \rightarrow \begin{array}{r} 190 \\ +\ 18 \\ \hline 208 \end{array}$$

Elementary Example 2

19 × 11

Step 1. Add 1 to the 19: $19 + 1 = 20$.

Step 2. Multiply the 20 by 10: $20 \times 10 = 200$.

Step 3. Multiply together the ones digits: $9 \times 1 = 9$.

Step 4. Add the 9 to the 200: $200 + 9 = 209$ (answer).

Thought Process Summary

$$\begin{array}{r} 19 \\ \times 11 \\ \hline \end{array} \rightarrow \begin{array}{r} 19 \\ +\ 1 \\ \hline 20 \end{array} \rightarrow \begin{array}{r} 20 \\ \times 10 \\ \hline 200 \end{array} \rightarrow \begin{array}{r} 9 \\ \times 1 \\ \hline 9 \end{array} \rightarrow \begin{array}{r} 200 \\ +\ 9 \\ \hline 209 \end{array}$$

Brain Builder

18 × 17

Step 1. Add 7 to the 18: 18 + 7 = 25.

Step 2. Multiply the 25 by 10: 25 × 10 = 250.

Step 3. Multiply together the ones digits: 8 × 7 = 56.

Step 4. Add the 56 to the 250: 250 + 56 = 306 (answer).

Thought Process Summary

$$\begin{array}{r} 18 \\ \times 17 \\ \hline \end{array} \rightarrow \begin{array}{r} 18 \\ +\ 7 \\ \hline 25 \end{array} \rightarrow \begin{array}{r} 25 \\ \times 10 \\ \hline 250 \end{array} \rightarrow \begin{array}{r} 8 \\ \times 7 \\ \hline 56 \end{array} \rightarrow \begin{array}{r} 250 \\ +\ 56 \\ \hline 306 \end{array}$$

Food for Thought: In Elementary Example 1, for example, we could have taken 16 + 3 instead of 13 + 6 to obtain the 19 in Step 1. It doesn't matter which ones digit is being added to which factor. Finally, would you like a shortcut for multiplying numbers in their 20s (e.g., 21 × 26)? Well, just apply Trick 27 as shown above, except that you'll need to double the amount obtained in Step 1. Try it and see for yourself!

Practice Problems

To solve these, you'll need to add, multiply, multiply, and add.

Elementary Exercises

1. 12 × 15 =
   [(12 + 5) × 10] + (2 × 5) = 180

2. 18 × 11 =

3. 16 × 16 =

4. 17 × 13 =

5. $14 \times 12 =$

6. $15 \times 19 =$

7. $11 \times 19 =$

8. $18 \times 14 =$

9. $12 \times 19 =$

10. $17 \times 14 =$

11. $14 \times 14 =$

12. $13 \times 16 =$

13. $18 \times 13 =$

14. $15 \times 18 =$

Brain Builders

1. $19 \times 19 =$

2. $17 \times 16 =$

3. $17 \times 19 =$

4. $18 \times 18 =$

5. $19 \times 16 =$

6. $18 \times 19 =$

7. $17 \times 18 =$

8. $17 \times 17 =$

(See solutions on page 207.)

Mathematical Curiosity

$$1 \times 1 = 1$$
$$11 \times 11 = 121$$
$$111 \times 111 = 12321$$
$$1111 \times 1111 = 1234321$$
$$11111 \times 11111 = 123454321$$

etc.

Trick 28: Squaring Any Number Ending in 1 or 9

Strategy: It's probably safe to assume that multiplying by a number that ends in zero is so much easier than multiplying by one that doesn't. As such, we are going to square numbers ending in 1 or 9 by multiplying together the two whole numbers on either side of the number being squared and then adding 1. Let's look at some step-by-step explanations below.

Elementary Example 1

21 × 21 (or 21^2)

Step 1. Multiply together the whole numbers on either side of 21: 22 × 20 = 440.

Step 2. Add 1 to the 440: 440 + 1 = 441 (answer).

Thought Process Summary

$$\begin{array}{r} 21 \\ \times\ 21 \\ \hline \end{array} \rightarrow \begin{array}{r} 22 \\ \times\ 20 \\ \hline 440 \end{array} \rightarrow \begin{array}{r} 440 \\ +\ \ 1 \\ \hline 441 \end{array}$$

Elementary Example 2

19 × 19 (or 19^2)

Step 1. Multiply together the whole numbers on either side of 19: 18 × 20 = 360.

Step 2. Add 1 to the 360: 360 + 1 = 361 (answer).

Thought Process Summary

$$\begin{array}{r} 19 \\ \times\ 19 \\ \hline \end{array} \rightarrow \begin{array}{r} 18 \\ \times\ 20 \\ \hline 360 \end{array} \rightarrow \begin{array}{r} 360 \\ +\ \ 1 \\ \hline 361 \end{array}$$

Brain Builder

81 × 81 (or 81^2)

Step 1. Multiply together the whole numbers on either side of 81: 82 × 80 = 6,560.

Step 2. Add 1 to the 6,560: 6,560 + 1 = 6,561 (answer).

Thought Process Summary

$$\begin{array}{r} 81 \\ \times 81 \\ \hline \end{array} \rightarrow \begin{array}{r} 82 \\ \times 80 \\ \hline 6{,}560 \end{array} \rightarrow \begin{array}{r} 6{,}560 \\ + \quad 1 \\ \hline 6{,}561 \end{array}$$

Food for Thought: The number to add in the last step (1) is easy to remember because you are multiplying together the two whole numbers that are **1** (or, technically, 1^2) away from the number being squared.

Practice Problems

Solve these problems by multiplying together the whole numbers on either side and adding 1.

Elementary Exercises

1. $31^2 = \mathbf{(32 \times 30) + 1 = 961}$
2. $29^2 =$
3. $51^2 =$
4. $39^2 =$
5. $49 \times 49 =$
6. $61 \times 61 =$
7. $19 \times 19 =$
8. $41 \times 41 =$
9. $21^2 =$
10. $59^2 =$
11. $61^2 =$
12. $49^2 =$
13. $29 \times 29 =$
14. $31 \times 31 =$

Brain Builders

1. $79^2 =$
2. $91^2 =$
3. $69^2 =$
4. $81^2 =$
5. $71 \times 71 =$
6. $89 \times 89 =$
7. $99 \times 99 =$
8. $101 \times 101 =$

(See solutions on page 208.)

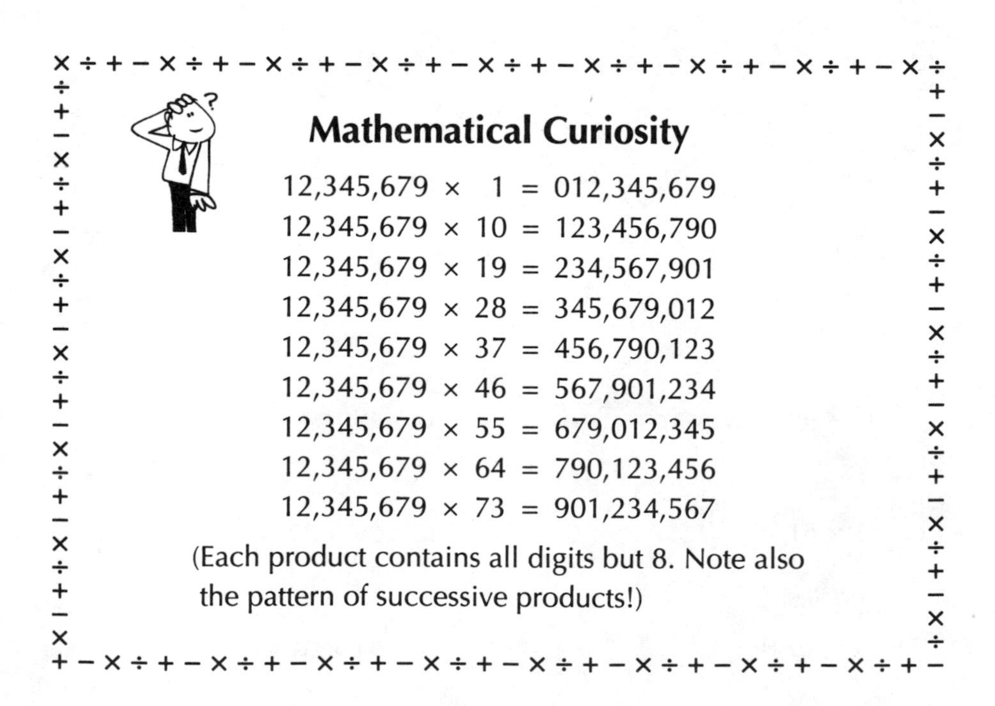

Mathematical Curiosity

12,345,679 × 1 = 012,345,679
12,345,679 × 10 = 123,456,790
12,345,679 × 19 = 234,567,901
12,345,679 × 28 = 345,679,012
12,345,679 × 37 = 456,790,123
12,345,679 × 46 = 567,901,234
12,345,679 × 55 = 679,012,345
12,345,679 × 64 = 790,123,456
12,345,679 × 73 = 901,234,567

(Each product contains all digits but 8. Note also the pattern of successive products!)

Trick 29: Squaring Any Number Ending in 2 or 8

Strategy: This trick is identical to Trick 28 with two exceptions. First, to square a number ending in 2 or 8, multiply together the numbers that are 2 above and 2 below the number being squared. Second, add 4 to the product. As you will see in the examples below, we will simplify the calculation by converting one of the multipliers to a multiple of 10.

Elementary Example 1

22 × 22 (or 22^2)

Step 1. Multiply together the numbers that are 2 above and 2 below 22: 24 × 20 = 480.

Step 2. Add 4 to the 480: 480 + 4 = 484 (answer).

Thought Process Summary

$$\begin{array}{r} 22 \\ \times\ 22 \\ \hline \end{array} \rightarrow \begin{array}{r} 24 \\ \times\ 20 \\ \hline 480 \end{array} \rightarrow \begin{array}{r} 480 \\ +\ \ 4 \\ \hline 484 \end{array}$$

Elementary Example 2

18 × 18 (or 18^2)

Step 1. Multiply together the numbers that are 2 above and 2 below 18: 16 × 20 = 320.

Step 2. Add 4 to the 320: 320 + 4 = 324 (answer).

Thought Process Summary

$$\begin{array}{r} 18 \\ \times\ 18 \\ \hline \end{array} \rightarrow \begin{array}{r} 16 \\ \times\ 20 \\ \hline 320 \end{array} \rightarrow \begin{array}{r} 320 \\ +\ \ 4 \\ \hline 324 \end{array}$$

Brain Builder

72 × 72 (or 72^2)

Step 1. Multiply together the numbers that are 2 above and 2 below 72: 74 × 70 = 5,180.

Step 2. Add 4 to the 5,180: 5,180 + 4 = 5,184 (answer).

Thought Process Summary

$$\begin{array}{r} 72 \\ \times\ 72 \\ \hline \end{array} \rightarrow \begin{array}{r} 74 \\ \times\ 70 \\ \hline 5{,}180 \end{array} \rightarrow \begin{array}{r} 5{,}180 \\ +\ \ \ \ 4 \\ \hline 5{,}184 \end{array}$$

Food for Thought: The number to add in the last step (4) is easy to remember—just square the distance you have moved above and below the number being squared in the previous step (i.e., 2^2 equals 4).

Practice Problems

Solve these problems by multiplying together the numbers that are 2 above and below and adding 4.

Elementary Exercises

1. 28 × 28 = **(26 × 30) + 4 = 784**
2. 32 × 32 =
3. 38 × 38 =
4. 52 × 52 =
5. 62^2 =
6. 48^2 =

7. $18^2 =$
8. $42^2 =$
9. $22 \times 22 =$
10. $58 \times 58 =$
11. $62 \times 62 =$
12. $48 \times 48 =$
13. $32^2 =$
14. $28^2 =$

Brain Builders

1. $82 \times 82 =$
2. $68 \times 68 =$
3. $92 \times 92 =$
4. $78 \times 78 =$
5. $72^2 =$
6. $88^2 =$
7. $98^2 =$
8. $102^2 =$

(See solutions on page 208.)

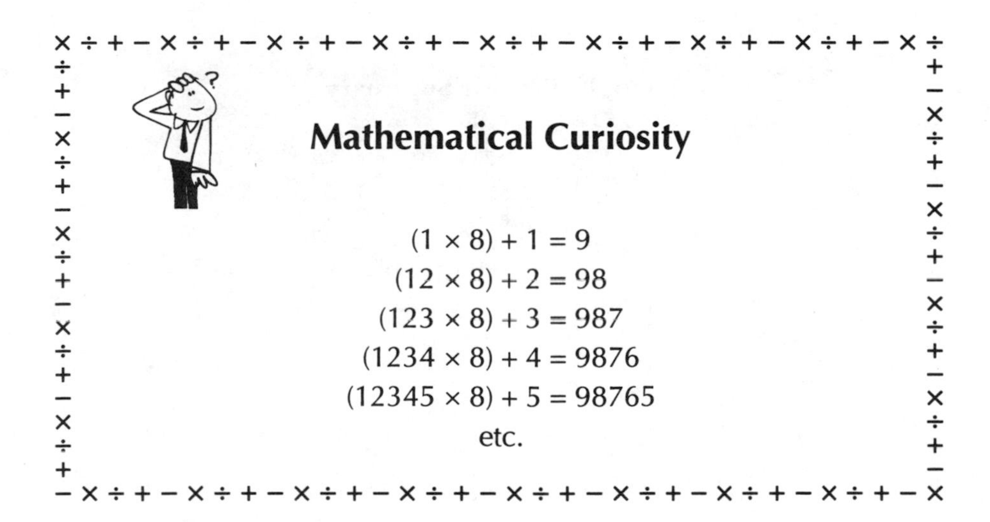

Trick 30: Squaring Any Number Ending in 3 or 7

Strategy: Perhaps you've already guessed the secret behind this trick. It's identical to Tricks 28 and 29, with two exceptions. First, to square a number ending in 3 or 7, multiply together the numbers that are 3 above and 3 below the number being squared. Second, add 9 to the product. Once again, as you will see in the examples below, we will simplify the calculation by converting one of the multipliers to a multiple of 10.

Elementary Example 1

13 × 13 (or 13^2)

Step 1. Multiply together the numbers that are 3 above and 3 below 13: $16 \times 10 = 160$.

Step 2. Add 9 to the 160: $160 + 9 = 169$ (answer).

Thought Process Summary

$$\begin{array}{r} 13 \\ \times\ 13 \\ \hline \end{array} \rightarrow \begin{array}{r} 16 \\ \times\ 10 \\ \hline 160 \end{array} \rightarrow \begin{array}{r} 160 \\ +\ \ 9 \\ \hline 169 \end{array}$$

Elementary Example 2

17 × 17 (or 17^2)

Step 1. Multiply together the numbers that are 3 above and 3 below 17: $14 \times 20 = 280$.

Step 2. Add 9 to the 280: $280 + 9 = 289$ (answer).

Thought Process Summary

$$\begin{array}{r} 17 \\ \times\ 17 \\ \hline \end{array} \rightarrow \begin{array}{r} 14 \\ \times\ 20 \\ \hline 280 \end{array} \rightarrow \begin{array}{r} 280 \\ +\ \ 9 \\ \hline 289 \end{array}$$

Brain Builder

83 × 83 (or 83^2)

Step 1. Multiply together the numbers that are 3 above and 3 below 83: 86 × 80 = 6,880.

Step 2. Add 9 to the 6,880: 6,880 + 9 = 6,889 (answer).

Thought Process Summary

$$\begin{array}{r} 83 \\ \times\ 83 \\ \hline \end{array} \rightarrow \begin{array}{r} 86 \\ \times\ 80 \\ \hline 6{,}880 \end{array} \rightarrow \begin{array}{r} 6{,}880 \\ +\quad 9 \\ \hline 6{,}889 \end{array}$$

Food for Thought: The number to add in the last step (9) is easy to remember—just square the distance you have moved above and below the number being squared in the previous step (i.e., $3^2 = 9$).

Practice Problems

Solve these problems by multiplying together the numbers that are 3 above and below and adding 9.

Elementary Exercises

1. 23^2 = **(26 × 20) + 9 = 529**
2. 27^2 =
3. 37^2 =
4. 53^2 =
5. 63 × 63 =
6. 47 × 47 =
7. 17 × 17 =
8. 43 × 43 =
9. 37 × 37 =
10. 33^2 =

11. $13^2 =$

12. $57^2 =$

13. $47^2 =$

14. $63^2 =$

Brain Builders

1. $83^2 =$
2. $67^2 =$
3. $93^2 =$
4. $77^2 =$
5. $73 \times 73 =$
6. $87 \times 87 =$
7. $97 \times 97 =$
8. $103 \times 103 =$

(See solutions on page 208.)

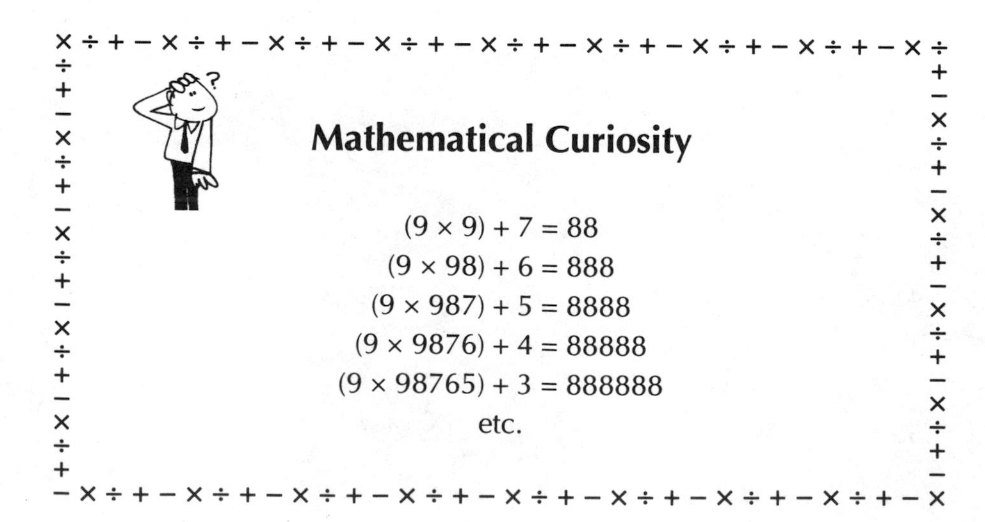

Trick 31: Multiplying by Multiples of 11

Strategy: This nifty trick involves multiplication by 11, 22, 33, and so forth. The secret is to multiply by the previous multiple of 10, then add 10%. As we have seen in several other tricks, converting one of the multipliers to a multiple of 10 simplifies the computation significantly. Let's now try a few out.

Elementary Example 1

7 × 33

Step 1. Determine the closest multiple of 10 under 33: 30.

Step 2. Multiply 7 by 30: $7 \times 30 = 210$.

Step 3. Calculate 10% of 210: $210 \times 10\% = 21$.

Step 4. Add the 210 and the 21: $210 + 21 = 231$ (answer).

Thought Process Summary

$$\begin{array}{r} 7 \\ \times\, 33 \\ \hline \end{array} \rightarrow \begin{array}{r} 7 \\ \times\, 30 \\ \hline 210 \end{array} \rightarrow \begin{array}{r} 210 \\ \times\, 10\% \\ \hline 21 \end{array} \rightarrow \begin{array}{r} 210 \\ +\, 21 \\ \hline 231 \end{array}$$

Elementary Example 2

8 × 55

Step 1. Determine the closest multiple of 10 under 55: 50.

Step 2. Multiply 8 by 50: $8 \times 50 = 400$.

Step 3. Calculate 10% of 400: $400 \times 10\% = 40$.

Step 4. Add the 400 and the 40: $400 + 40 = 440$ (answer).

Thought Process Summary

$$\begin{array}{r} 8 \\ \times\ 55 \\ \hline \end{array} \rightarrow \begin{array}{r} 8 \\ \times\ 50 \\ \hline 400 \end{array} \rightarrow \begin{array}{r} 400 \\ \times\ 10\% \\ \hline 40 \end{array} \rightarrow \begin{array}{r} 400 \\ +\ 40 \\ \hline 440 \end{array}$$

Brain Builder

45 × 66

Step 1. Determine the closest multiple of 10 under 66: 60.
Step 2. Multiply 45 by 60: 45 × 60 = 2,700.
Step 3. Calculate 10% of 2,700: 2,700 × 10% = 270.
Step 4. Add the 2,700 and the 270: 2,700 + 270 = 2,970 (answer).

Thought Process Summary

$$\begin{array}{r} 45 \\ \times\ 66 \\ \hline \end{array} \rightarrow \begin{array}{r} 45 \\ \times\ 60 \\ \hline 2{,}700 \end{array} \rightarrow \begin{array}{r} 2{,}700 \\ \times\ \ 10\% \\ \hline 270 \end{array} \rightarrow \begin{array}{r} 2{,}700 \\ +\ \ 270 \\ \hline 2{,}970 \end{array}$$

Food for Thought: Remember that multiplication by 10% is easy and the same as dividing by 10 (or taking $\frac{1}{10}$). Just move the decimal point (of the number you're multiplying) one place to the left, or lop off a right-hand zero. For example, 10% of 314 is 31.4, whereas 10% of 850 is 85.

Practice Problems

Multiply by the previous multiple of 10, add 10%, and you're in business!

Elementary Exercises

1. 9 × 22 = **(9 × 20) + 18 = 198**
2. 6 × 77 =
3. 7 × 44 =
4. 8 × 99 =
5. 6 × 44 =
6. 5 × 66 =

7. $8 \times 33 =$
8. $4 \times 88 =$
9. $7 \times 55 =$
10. $8 \times 22 =$
11. $9 \times 44 =$
12. $5 \times 77 =$
13. $6 \times 99 =$
14. $7 \times 66 =$

Brain Builders

1. $14 \times 55 =$
2. $65 \times 22 =$
3. $30 \times 77 =$
4. $17 \times 33 =$
5. $15 \times 88 =$
6. $35 \times 44 =$
7. $40 \times 99 =$
8. $12 \times 66 =$

(See solutions on page 209.)

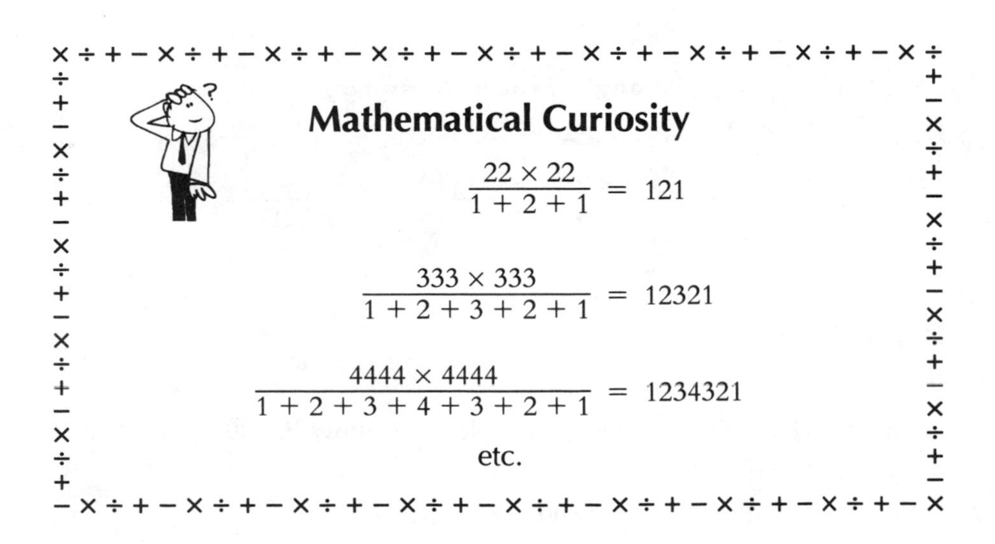

Mathematical Curiosity

$$\frac{22 \times 22}{1 + 2 + 1} = 121$$

$$\frac{333 \times 333}{1 + 2 + 3 + 2 + 1} = 12321$$

$$\frac{4444 \times 4444}{1 + 2 + 3 + 4 + 3 + 2 + 1} = 1234321$$

etc.

Trick 32: Multiplying by Multiples of 9

Strategy: This trick involves multiplication by 9, 18, 27, and so forth. The secret is to multiply by the next multiple of 10, then subtract 10%. As you will see in the examples below, this trick is very similar to Trick 31.

Elementary Example 1

8 × 27

Step 1. Determine the closest multiple of 10 above 27: 30.

Step 2. Multiply 8 by 30: 8 × 30 = 240.

Step 3. Calculate 10% of 240: 240 × 10% = 24.

Step 4. Subtract the 24 from the 240: 240 − 24 = 216 (answer).

Thought Process Summary

$$\begin{array}{r} 8 \\ \times\ 27 \\ \hline \end{array} \rightarrow \begin{array}{r} 8 \\ \times\ 30 \\ \hline 240 \end{array} \rightarrow \begin{array}{r} 240 \\ \times\ 10\% \\ \hline 24 \end{array} \rightarrow \begin{array}{r} 240 \\ -\ 24 \\ \hline 216 \end{array}$$

Elementary Example 2

7 × 18

Step 1. Determine the closest multiple of 10 above 18: 20.

Step 2. Multiply 7 by 20: 7 × 20 = 140.

Step 3. Calculate 10% of 140: 140 × 10% = 14.

Step 4. Subtract the 14 from the 140: 140 − 14 = 126 (answer).

Thought Process Summary

$$\begin{array}{r} 7 \\ \times\ 18 \\ \hline \end{array} \rightarrow \begin{array}{r} 7 \\ \times\ 20 \\ \hline 140 \end{array} \rightarrow \begin{array}{r} 140 \\ \times\ 10\% \\ \hline 14 \end{array} \rightarrow \begin{array}{r} 140 \\ -\ 14 \\ \hline 126 \end{array}$$

Brain Builder

35 × 72

Step 1. Determine the closest multiple of 10 above 72: 80.

Step 2. Multiply 35 by 80: 35 × 80 = 2,800.

Step 3. Calculate 10% of 2,800: 2,800 × 10% = 280.

Step 4. Subtract the 280 from the 2,800: 2,800 − 280 = 2,520 (answer).

Thought Process Summary

$$\begin{array}{r} 35 \\ \times\ 72 \\ \hline \end{array} \rightarrow \begin{array}{r} 35 \\ \times\ 80 \\ \hline 2{,}800 \end{array} \rightarrow \begin{array}{r} 2{,}800 \\ \times\ \ 10\% \\ \hline 280 \end{array} \rightarrow \begin{array}{r} 2{,}800 \\ -\ \ 280 \\ \hline 2{,}520 \end{array}$$

Food for Thought: Trick 32 is a tad more difficult to apply than Trick 31 because we must subtract the 10% rather than add it, and subtraction is usually more difficult than addition. Because the two tricks are so similar, however, you should make sure you've learned which trick converts to the next multiple of 10 and which one to the previous multiple of 10, as well as which one adds the 10% and which one subtracts it.

Practice Problems

Multiply by the next multiple of 10, subtract 10%, and you're home free!

Elementary Exercises

1. $7 \times 36 =$ **$(7 \times 40) - 28 = 252$**
2. $5 \times 63 =$
3. $9 \times 18 =$
4. $8 \times 81 =$
5. $6 \times 27 =$
6. $17 \times 9 =$
7. $7 \times 72 =$
8. $9 \times 45 =$
9. $6 \times 54 =$
10. $7 \times 18 =$
11. $9 \times 36 =$
12. $8 \times 27 =$
13. $19 \times 9 =$
14. $8 \times 45 =$

Brain Builders

1. $13 \times 54 =$
2. $35 \times 36 =$
3. $15 \times 72 =$
4. $67 \times 9 =$
5. $30 \times 81 =$
6. $75 \times 18 =$
7. $12 \times 45 =$
8. $25 \times 27 =$

(See solutions on page 209.)

Number Potpourri 13

Quick Check

Here's how to quickly check your answer when multiplying by 9, 99, 999, and so forth. All you have to do is add up the digits of the product. If that sum can be evenly divided by 9 (i.e., if it is 9, 18, 27, etc.), your answer is *probably* correct. For example, the calculation $99 \times 426 = 42{,}174$ is probably correct because the digit-sum of the product $(4 + 2 + 1 + 7 + 4)$ is 18. If the sum of the product's digits *cannot* be divided evenly by 9, then the product is *definitely* incorrect. So the calculation $64 \times 999 = 63{,}836$ is definitely incorrect because the product's digit-sum $(6 + 3 + 8 + 3 + 6)$ is 26, which cannot be evenly divided by 9. This checking technique will also work if the digit-sum of either factor is 9, such as 432, or a multiple of 9, such as 648.

Trick 33: Squaring Any Number Ending in 6

Prerequisite: Before you proceed with this trick, please turn to Appendix A ("Tricks from *Rapid Math Tricks and Tips* That You Need to Know for This Book"). Specifically, you will need to know how to square a number ending in 5 (page 193).

Strategy: To square a number ending in 6, first square the number that is one less, so that you're squaring a number ending in 5. Use the "squaring any number ending in 5" trick for this purpose. Then add the number that is one less than the original number you're squaring and the number you're squaring. For example, $16^2 = 15^2 + 15 + 16$, which equals $225 + 15 + 16$, or 256.

Elementary Example 1

16 × 16 (or 16^2)

Step 1. Square the number one below 16: $15^2 = 225$.

Step 2. Add 15 and 16 to the 225: $225 + 15 + 16 = 256$ (answer).

Thought Process Summary

$16^2 \rightarrow 15^2 + 15 + 16 \rightarrow 225 + 15 + 16 \rightarrow 256$

Elementary Example 2

26 × 26 (or 26^2)

Step 1. Square the number one below 26: $25^2 = 625$.

Step 2. Add 25 and 26 to the 625: $625 + 25 + 26 = 676$ (answer).

Thought Process Summary

$26^2 \rightarrow 25^2 + 25 + 26 \rightarrow 625 + 25 + 26 \rightarrow 676$

Brain Builder

76 × 76 (or 76^2)

Step 1. Square the number one below 76: $75^2 = 5{,}625$.

Step 2. Add 75 and 76 to the 5,625: $5{,}625 + 75 + 76 = 5{,}776$ (answer).

Thought Process Summary

$76^2 \rightarrow 75^2 + 75 + 76 \rightarrow 5{,}625 + 75 + 76 \rightarrow 5{,}776$

Practice Problems

Make sure you've mastered the squaring of numbers ending in 5 before proceeding with these exercises.

Elementary Exercises

1. $36^2 = \mathbf{35^2 + 35 + 36 = 1{,}296}$
2. $56^2 =$
3. $16^2 =$
4. $46^2 =$
5. $26^2 =$
6. $16 \times 16 =$
7. $46 \times 46 =$
8. $36 \times 36 =$
9. $56 \times 56 =$
10. $26 \times 26 =$

Brain Builders

1. $86 \times 86 =$
2. $66 \times 66 =$
3. $96 \times 96 =$
4. $76 \times 76 =$
5. $66^2 =$
6. $86^2 =$
7. $76^2 =$
8. $106^2 =$

(See solutions on page 209.)

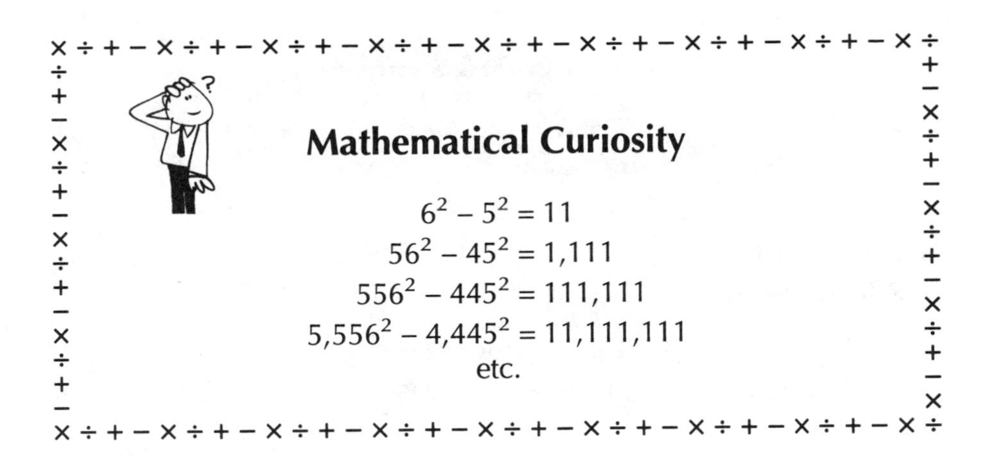

Trick 34: Squaring Any Number Ending in 4

Prerequisite: If you just completed Trick 33, then you're ready for Trick 34. However, if you did not, you should turn to page 193 of Appendix A to learn how to square any number ending in 5. You will need that knowledge to be able to square numbers ending in 4.

Strategy: The technique involved is different from that of Trick 33 in that you subtract numbers and work your way down rather than up. For example, to square the number 14, you would take $15^2 - 15 - 14$, which equals $225 - 15 - 14$, or 196. (Notice how we again make use of the "squaring any number ending in 5" trick.) In the examples below, notice the pattern of the numbers involved.

Elementary Example 1

14 × 14 (or 14^2)

Step 1. Square the number one above 14: $15^2 = 225$.

Step 2. Subtract 15 and 14 from the 225: $225 - 15 - 14 = 196$ (answer).

Thought Process Summary

$14^2 \rightarrow 15^2 - 15 - 14 \rightarrow 225 - 15 - 14 \rightarrow 196$

Elementary Example 2

24 × 24 (or 24^2)

Step 1. Square the number one above 24: $25^2 = 625$.

Step 2. Subtract 25 and 24 from the 625: $625 - 25 - 24 = 576$ (answer).

Thought Process Summary

$24^2 \rightarrow 25^2 - 25 - 24 \rightarrow 625 - 25 - 24 \rightarrow 576$

Brain Builder

84 × 84 (or 84^2)

Step 1. Square the number one above 84: $85^2 = 7{,}225$.

Step 2. Subtract 85 and 84 from the 7,225: $7{,}225 - 85 - 84 = 7{,}056$ (answer).

Thought Process Summary

$84^2 \rightarrow 85^2 - 85 - 84 \rightarrow 7{,}225 - 85 - 84 \rightarrow 7{,}056$

Food for Thought: The numbers subtracted in Step 2 are one more than the number being squared and the number being squared (85 and 84 in our Brain Builder example). Even though Tricks 33 and 34 are fairly similar to each other, Trick 34 requires a bit more concentration because of the subtraction involved.

Practice Problems

Begin these exercises by squaring the number that is one whole number above.

Elementary Exercises

1. $34 \times 34 =$
   $35^2 - 35 - 34 = 1{,}156$
2. $54 \times 54 =$
3. $14 \times 14 =$
4. $44 \times 44 =$
5. $24 \times 24 =$
6. $14^2 =$
7. $44^2 =$
8. $34^2 =$
9. $54^2 =$
10. $24^2 =$

Brain Builders

1. $84^2 =$
2. $64^2 =$
3. $94^2 =$
4. $74^2 =$
5. $64 \times 64 =$
6. $74 \times 74 =$
7. $84 \times 84 =$
8. $104 \times 104 =$

(See solutions on page 210.)

Trick 35: Place-Value Multiplication: Four-Digit by One-Digit

Strategy: This trick is similar to Tricks 16 and 17 because it focuses on what each digit represents within a number. For example, the number 1,234 represents (1 × 1,000) + (2 × 100) + (3 × 10) + (4 × 1). Let's now apply place-value multiplication to some four-digit by one-digit calculations.

Elementary Example 1

2,534 × 7

Step 1. Recast 2,534, emphasizing place value: 2,000 + 500 + 30 + 4.

Step 2. Multiply 7 by 2,000: 7 × 2,000 = 14,000.

Step 3. Multiply 7 by 500: 7 × 500 = 3,500.

Step 4. Multiply 7 by 30: 7 × 30 = 210.

Step 5. Multiply 7 by 4: 7 × 4 = 28.

Step 6. Add the above products: 14,000 + 3,500 + 210 + 28= 17,738 (answer).

Thought Process Summary

$$\begin{array}{r} 2{,}534 \\ \times \quad 7 \\ \hline \end{array} \rightarrow \begin{array}{r} (2{,}000 + 500 + 30 + 4) \\ \times \qquad\qquad\qquad\qquad 7 \\ \hline \end{array} \rightarrow$$

$$\begin{array}{r} 2{,}000 \\ \times \quad 7 \\ \hline 14{,}000 \end{array} + \begin{array}{r} 500 \\ \times \ 7 \\ \hline 3{,}500 \end{array} + \begin{array}{r} 30 \\ \times 7 \\ \hline 210 \end{array} + \begin{array}{r} 4 \\ \times 7 \\ \hline 28 \end{array} = 17{,}738$$

Elementary Example 2

4,051 × 6

Step 1. Recast 4,051, emphasizing place value: 4,000 + 50 + 1.

Step 2. Multiply 6 by 4,000: 6 × 4,000 = 24,000.

Step 3. Multiply 6 by 50: 6 × 50 = 300.

Step 4. Multiply 6 by 1: 6 × 1 = 6.

Step 5. Add the above products: 24,000 + 300 + 6 = 24,306 (answer).

Thought Process Summary

$$\begin{array}{r} 4{,}051 \\ \times \quad 6 \\ \hline \end{array} \rightarrow \begin{array}{r} (4{,}000 + 50 + 1) \\ \times \qquad\qquad 6 \\ \hline \end{array} \rightarrow$$

$$\begin{array}{r} 4{,}000 \\ \times \quad 6 \\ \hline 24{,}000 \end{array} + \begin{array}{r} 50 \\ \times 6 \\ \hline 300 \end{array} + \begin{array}{r} 1 \\ \times 6 \\ \hline 6 \end{array} = 24{,}306$$

Brain Builder

8,367 × 9

Step 1. Recast 8,367, emphasizing place value: 8,000 + 300 + 60 + 7.

Step 2. Multiply 9 by 8,000: 9 × 8,000 = 72,000.

Step 3. Multiply 9 by 300: 9 × 300 = 2,700.

Step 4. Multiply 9 by 60: 9 × 60 = 540.

Step 5. Multiply 9 by 7: 9 × 7 = 63.

Step 6. Add the above products: 72,000 + 2,700 + 540 + 63 = 75,303 (answer).

Thought Process Summary

$$\begin{array}{r} 8{,}367 \\ \times \quad 9 \\ \hline \end{array} \rightarrow \begin{array}{r} (8{,}000 + 300 + 60 + 7) \\ \times \qquad\qquad\qquad 9 \\ \hline \end{array} \rightarrow$$

$$\begin{array}{r} 8{,}000 \\ \times \quad 9 \\ \hline 72{,}000 \end{array} + \begin{array}{r} 300 \\ \times \ 9 \\ \hline 2{,}700 \end{array} + \begin{array}{r} 60 \\ \times 9 \\ \hline 540 \end{array} + \begin{array}{r} 7 \\ \times 9 \\ \hline 63 \end{array} = 75{,}303$$

Food for Thought: This trick obviously requires much practice and concentration. The advantage, however, is that you can perform the entire computation (from left to right) in your head.

Practice Problems

Concentrate hard as you perform left-to-right multiplication.

Elementary Exercises

1. $4{,}132 \times 5$
   20,660
   (20,000 + 500 + 150 + 10)

2. $3{,}250 \times 8$

3. $1{,}524 \times 7$

4. $2{,}205 \times 9$

5. $3{,}540 \times 6$

6. $4{,}053 \times 4$

7. $5{,}133 \times 5$

8. $1{,}234 \times 8$

9. $4{,}321 \times 3$

10. $3{,}232 \times 7$

11. $4{,}204 \times 6$

12. $2{,}544 \times 5$

13. $3{,}512 \times 9$

14. $1{,}514 \times 8$

Brain Builders

1. $\begin{array}{r} 7{,}345 \\ \times \quad 3 \\ \hline \end{array}$

2. $\begin{array}{r} 3{,}906 \\ \times \quad 6 \\ \hline \end{array}$

3. $\begin{array}{r} 9{,}822 \\ \times \quad 7 \\ \hline \end{array}$

4. $\begin{array}{r} 6{,}053 \\ \times \quad 5 \\ \hline \end{array}$

5. $\begin{array}{r} 8{,}251 \\ \times \quad 4 \\ \hline \end{array}$

6. $\begin{array}{r} 4{,}776 \\ \times \quad 8 \\ \hline \end{array}$

7. $\begin{array}{r} 1{,}089 \\ \times \quad 6 \\ \hline \end{array}$

8. $\begin{array}{r} 7{,}465 \\ \times \quad 7 \\ \hline \end{array}$

(See solutions on page 210.)

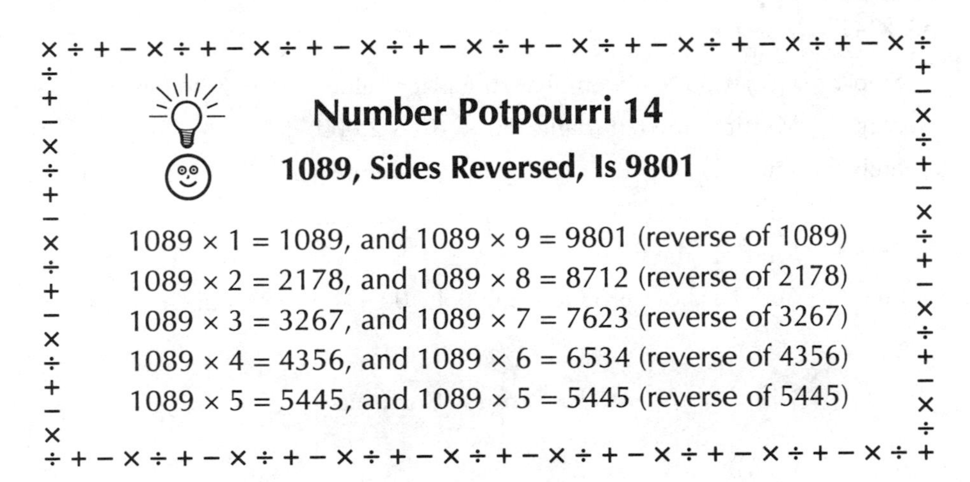

Number Potpourri 14

1089, Sides Reversed, Is 9801

1089 × 1 = 1089, and 1089 × 9 = 9801 (reverse of 1089)
1089 × 2 = 2178, and 1089 × 8 = 8712 (reverse of 2178)
1089 × 3 = 3267, and 1089 × 7 = 7623 (reverse of 3267)
1089 × 4 = 4356, and 1089 × 6 = 6534 (reverse of 4356)
1089 × 5 = 5445, and 1089 × 5 = 5445 (reverse of 5445)

Trick 36: Place-Value Multiplication: Two-Digit by Two-Digit

Strategy: This trick requires four multiplications, as you will see below. As with Trick 35, the significance of a digit within a number is vital. Remember, for example, that a tens digit equals that digit times 10.

Elementary Example 1

12 × 34

Step 1. Recast 12 × 34, emphasizing place value: (10 + 2) × (30 + 4).
Step 2. Multiply the tens digits: 10 × 30 = 300.
Step 3. Multiply one tens digit by the other ones digit: 10 × 4 = 40.
Step 4. Multiply together the other tens and ones digits: 30 × 2 = 60.
Step 5. Multiply the ones digits: 2 × 4 = 8.
Step 6. Add the above products: 300 + 40 + 60 + 8 = 408 (answer).

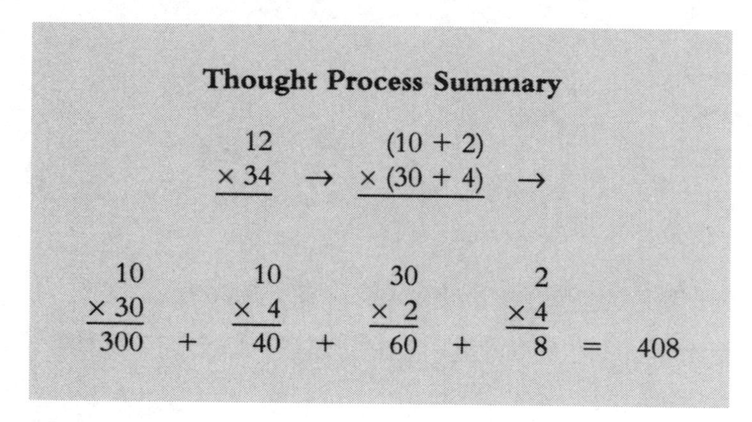

Elementary Example 2

53 × 42

Step 1. Recast 53 × 42, emphasizing place value: (50 + 3) × (40 + 2).
Step 2. Multiply the tens digits: 50 × 40 = 2,000.
Step 3. Multiply one tens digit by the other ones digit: 50 × 2 = 100.
Step 4. Multiply together the other tens and ones digits: 40 × 3 = 120.
Step 5. Multiply the ones digits: 3 × 2 = 6.
Step 6. Add the above products: 2,000 + 100 + 120 + 6 = 2,226 (answer).

Thought Process Summary

$$\begin{array}{r}53\\ \times\ 42\\ \hline\end{array} \rightarrow \begin{array}{r}(50+3)\\ \times\ (40+2)\\ \hline\end{array} \rightarrow$$

$$\begin{array}{r}50\\ \times\ 40\\ \hline 2{,}000\end{array} + \begin{array}{r}50\\ \times\ 2\\ \hline 100\end{array} + \begin{array}{r}40\\ \times\ 3\\ \hline 120\end{array} + \begin{array}{r}3\\ \times\ 2\\ \hline 6\end{array} = 2{,}226$$

Brain Builder

73 × 48

Step 1. Recast 73 × 48, emphasizing place value: (70 + 3) × (40 + 8).
Step 2. Multiply the tens digits: 70 × 40 = 2,800.
Step 3. Multiply one tens digit by the other ones digit: 70 × 8 = 560.
Step 4. Multiply together the other tens and ones digits: 40 × 3 = 120.
Step 5. Multiply the ones digits: 3 × 8 = 24.
Step 6. Add the above products: 2,800 + 560 + 120 + 24 = 3,504 (answer).

Thought Process Summary

$$\begin{array}{r} 73 \\ \times\ 48 \\ \hline \end{array} \rightarrow \begin{array}{r} (70 + 3) \\ \times\ (40 + 8) \\ \hline \end{array} \rightarrow$$

$$\begin{array}{r} 70 \\ \times\ 40 \\ \hline 2{,}800 \end{array} + \begin{array}{r} 70 \\ \times\ 8 \\ \hline 560 \end{array} + \begin{array}{r} 40 \\ \times\ 3 \\ \hline 120 \end{array} + \begin{array}{r} 3 \\ \times\ 8 \\ \hline 24 \end{array} = 3{,}504$$

Food for Thought: The possibilities for place-value multiplication are limitless, and it is a skill that is really impressive when mastered. We have limited the size of the calculations, however, to those that realistically can be accomplished with a reasonable amount of practice.

Practice Problems

Do all the work in your head, and just write down the answer!

Elementary Exercises

1. $\begin{array}{r} 23 \\ \times\ 14 \\ \hline \mathbf{322} \end{array}$
   (200 + 80 + 30 + 12)

2. $\begin{array}{r} 52 \\ \times\ 43 \\ \hline \end{array}$

3. $\begin{array}{r} 31 \\ \times\ 54 \\ \hline \end{array}$

4. $\begin{array}{r} 44 \\ \times\ 21 \\ \hline \end{array}$

5. $\begin{array}{r} 52 \\ \times\ 13 \\ \hline \end{array}$

6. $\begin{array}{r} 24 \\ \times\ 52 \\ \hline \end{array}$

7. 33×22

8. 15×34

9. 24×51

10. 14×34

11. 42×43

12. 25×35

13. 22×45

14. 55×33

Brain Builders

1. 73×46

2. 28×94

3. 47×61

4. 88×23

5. 52×39

6. 42×87

7. 18×66

8. 37×59

(See solutions on page 210.)

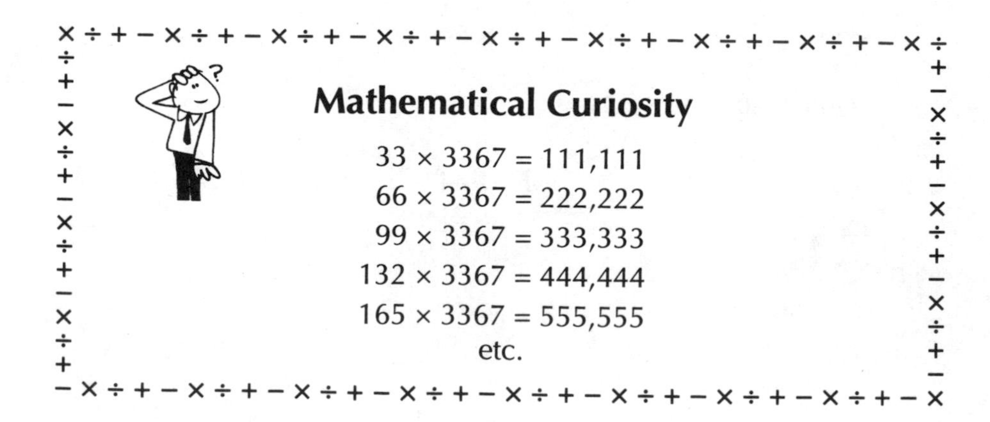

Trick 37: Squaring Numbers between 90 and 100

Strategy: This trick will work for calculations such as 97^2 and 91^2. The key is to first find the distance the number being squared is from 100. The left half of the answer will be the number being squared minus its distance from 100. The right half will equal the distance squared. (Squared distances of 1, 2, and 3 are written as 01, 04, and 09, respectively.) Let's look at some illustrations below.

Elementary Example 1

98 × 98 (or 98^2)

Step 1. Determine the distance 98 is from 100: 2.

Step 2. Subtract the 2 from 98: $98 - 2 = 96$ (left half of answer).

Step 3. Square the 2: $2^2 = 04$ (right half of answer).

Step 4. Combine the halves: 9,604 (answer).

Thought Process Summary

$$\begin{array}{r} 98 \\ \times\, 98 \\ \hline \end{array} \rightarrow \begin{array}{r} 100 \\ -\ 98 \\ \hline 2 \end{array} \rightarrow \begin{array}{r} 98 \\ -\ 2 \\ \hline 96 \end{array} \rightarrow \begin{array}{r} 2 \\ \times\, 2 \\ \hline 04 \end{array} \rightarrow 9{,}604$$

Elementary Example 2

93 × 93 (or 93^2)

Step 1. Determine the distance 93 is from 100: 7.

Step 2. Subtract the 7 from 93: $93 - 7 = 86$ (left half of answer).

Step 3. Square the 7: $7^2 = 49$ (right half of answer).

Step 4. Combine the halves: 8,649 (answer).

Thought Process Summary

$$\begin{array}{r} 93 \\ \times\ 93 \\ \hline \end{array} \rightarrow \begin{array}{r} 100 \\ -\ 93 \\ \hline 7 \end{array} \rightarrow \begin{array}{r} 93 \\ -\ 7 \\ \hline 86 \end{array} \rightarrow \begin{array}{r} 7 \\ \times\ 7 \\ \hline 49 \end{array} \rightarrow 8{,}649$$

Time Out: This trick will also work nicely when squaring numbers between 990 and 1,000, except that you must insert a zero into the hundreds place. Let's try it!

Brain Builder

996 × 996 (or 996^2)

Step 1. Determine the distance 996 is from 1,000: 4.

Step 2. Subtract the 4 from 996: 996 − 4 = 992 (left half of answer).

Step 3. Square the 4: 4^2 = 016 (right half of answer).

Step 4. Combine the halves: 992,016 (answer).

Thought Process Summary

$$\begin{array}{r} 996 \\ \times\ 996 \\ \hline \end{array} \rightarrow \begin{array}{r} 1{,}000 \\ -\ 996 \\ \hline 4 \end{array} \rightarrow \begin{array}{r} 996 \\ -\ 4 \\ \hline 992 \end{array} \rightarrow \begin{array}{r} 4 \\ \times\ 4 \\ \hline 016 \end{array} \rightarrow 992{,}016$$

Food for Thought: Notice in our Brain Builder example that the answer to 4 × 4 is written as 016 (i.e., a zero was inserted into the hundreds place). Trick 37 will work for squaring numbers 90 and under as well. However, those calculations are far more difficult to execute because they require carrying.

Practice Problems

Begin these problems by determining the distance from 100.

Elementary Exercises

1. 96 × 96 =
   Combine (96 − 4) and (4 × 4) = 9,216
2. 99 × 99 =
3. 95 × 95 =
4. 91 × 91 =

5. $97 \times 97 =$
6. $93 \times 93 =$
7. $98 \times 98 =$
8. $92 \times 92 =$
9. $94 \times 94 =$
10. $99^2 =$
11. $91^2 =$
12. $96^2 =$
13. $92^2 =$
14. $95^2 =$

Begin these problems by determining the distance from 1,000.

Brain Builders

1. $997 \times 997 =$
2. $994 \times 994 =$
3. $991 \times 991 =$
4. $999 \times 999 =$
5. $993^2 =$
6. $995^2 =$
7. $992^2 =$
8. $996^2 =$

(See solutions on page 211.)

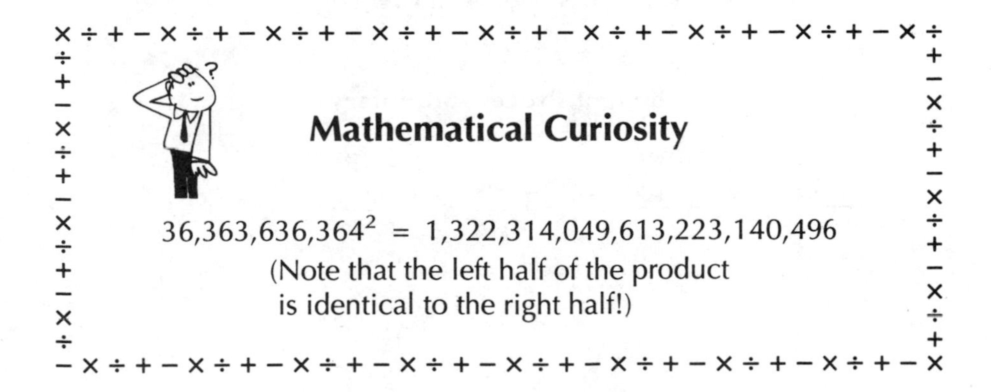

Trick 38: Squaring Numbers between 100 and 110

Strategy: This trick will work for calculations such as 102^2 and 107^2, and is executed in a manner similar to that of Trick 37. First, find the distance between the number being squared and 100 (that's easy). The left-hand portion of the answer will equal the number being squared plus its distance from 100. The right-hand portion will equal the distance squared. (Squared distances of 1, 2, and 3 are written as 01, 04, and 09, respectively.) As you'll see below, this trick is not nearly as difficult to execute as it sounds.

Elementary Example 1

103 × 103 (or 103^2)

Step 1. Determine the distance 103 is from 100: 3.

Step 2. Add the 3 to 103: 103 + 3 = 106 (left portion of answer).

Step 3. Square the 3: $3^2 = 09$ (right portion of answer).

Step 4. Combine the portions: 10,609 (answer).

Thought Process Summary

$$\begin{array}{r} 103 \\ \times\ 103 \\ \hline \end{array} \rightarrow \begin{array}{r} 103 \\ -\ 100 \\ \hline 3 \end{array} \rightarrow \begin{array}{r} 103 \\ +\ \ \ 3 \\ \hline 106 \end{array} \rightarrow \begin{array}{r} 3 \\ \times\ 3 \\ \hline 09 \end{array} \rightarrow 10{,}609$$

Elementary Example 2

108 × 108 (or 108^2)

Step 1. Determine the distance 108 is from 100: 8.

Step 2. Add the 8 to 108: 108 + 8 = 116 (left portion of answer).

Step 3. Square the 8: $8^2 = 64$ (right portion of answer).

Step 4. Combine the portions: 11,664 (answer).

Thought Process Summary

$$\begin{array}{r} 108 \\ \times\ 108 \\ \hline \end{array} \rightarrow \begin{array}{r} 108 \\ -\ 100 \\ \hline 8 \end{array} \rightarrow \begin{array}{r} 108 \\ +\ \ \ 8 \\ \hline 116 \end{array} \rightarrow \begin{array}{r} 8 \\ \times\ 8 \\ \hline 64 \end{array} \rightarrow 11{,}664$$

Time Out: This trick will also work nicely when squaring numbers between 1,000 and 1,010, except that you must insert a zero into the hundreds place. Let's try it!

Brain Builder

1,009 × 1,009 (or 1,009^2)

Step 1. Determine the distance 1,009 is from 1,000: 9.

Step 2. Add the 9 to 1,009: 1,009 + 9 = 1,018 (left portion of answer).

Step 3. Square the 9: 9^2 = 081 (right portion of answer).

Step 4. Combine the portions: 1,018,081 (answer).

Thought Process Summary

$$\begin{array}{r} 1{,}009 \\ \times\ 1{,}009 \\ \hline \end{array} \rightarrow \begin{array}{r} 1{,}009 \\ -\ 1{,}000 \\ \hline 9 \end{array} \rightarrow \begin{array}{r} 1{,}009 \\ +\ \quad 9 \\ \hline 1{,}018 \end{array} \rightarrow \begin{array}{r} 9 \\ \times\ 9 \\ \hline 081 \end{array} \rightarrow 1{,}018{,}081$$

Food for Thought: Notice in our Brain Builder example that the answer to 9 × 9 is written as 081 (i.e., a zero was inserted into the hundreds place). Trick 38 will work for squaring numbers 110 and over, as well. However, those calculations are far more difficult to execute because they require carrying.

Practice Problems

Make sure to add the distances from 100 when working these exercises.

Elementary Exercises

1. 104 × 104 =
   Combine (104 + 4)
   and (4 × 4) = 10,816

2. 101 × 101 =

3. 105 × 105 =

4. 109 × 109 =

5. 103 × 103 =

6. 107 × 107 =

7. 102 × 102 =

8. 108 × 108 =

9. $106 \times 106 =$

10. $101^2 =$

11. $109^2 =$

12. $104^2 =$

13. $108^2 =$

14. $105^2 =$

Add the distances from 1,000 when working these exercises.

Brain Builders

1. $1{,}003 \times 1{,}003 =$
2. $1{,}006 \times 1{,}006 =$
3. $1{,}009 \times 1{,}009 =$
4. $1{,}001 \times 1{,}001 =$
5. $1{,}007 \times 1{,}007 =$
6. $1{,}005 \times 1{,}005 =$
7. $1{,}008 \times 1{,}008 =$
8. $1{,}004 \times 1{,}004 =$

(See solutions on page 211.)

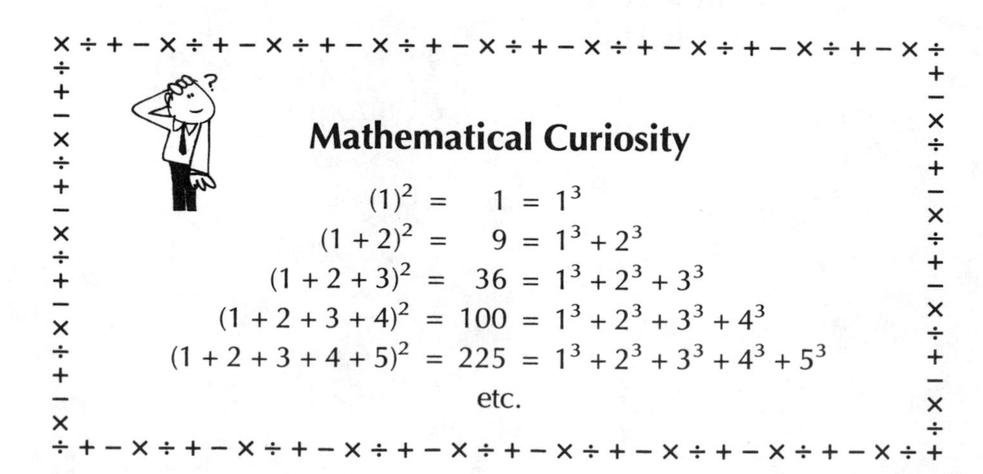

Trick 39: Multiplying Numbers Just under 100

Strategy: This trick will work for calculations such as 91 × 98 and 88 × 97. The secret is to determine each number's distance from 100, subtract from either number the distance from 100 of the other number, then multiply the two distances. One-digit products must be written with a zero to the left (e.g., 7 would be written as 07). As you will see below, Trick 39 bears a resemblance to Tricks 37 and 38.

Elementary Example 1

98 × 97

Step 1. Determine the distance 98 is from 100: 2.

Step 2. Determine the distance 97 is from 100: 3.

Step 3. Subtract 2 from 97 or 3 from 98: 97 − 2 (or 98 − 3) = 95 (left half of answer).

Step 4. Multiply together the two distances: 2 × 3 = 06 (right half of answer).

Step 5. Combine the results of Steps 3 and 4: 9,506 (answer).

Thought Process Summary

		Distance								
98	→	2		98		97		2		
× 97	→	3	→	− 3	or	− 2	→	× 3	→	9,506
				95		95		06		

Elementary Example 2

88 × 96

Step 1. Determine the distance 88 is from 100: 12.

Step 2. Determine the distance 96 is from 100: 4.

Step 3. Subtract 12 from 96 or 4 from 88: $96 - 12$ (or $88 - 4$)$= 84$ (left half of answer).

Step 4. Multiply together the two distances: $12 \times 4 = 48$ (right half of answer).

Step 5. Combine the results of Steps 3 and 4: 8,448 (answer).

Thought Process Summary

$$\begin{array}{rcrcrcrcrcl} & & \text{Distance} & & & & & & & & \\ 88 & \rightarrow & 12 & & 88 & & 96 & & 12 & & \\ \underline{\times\ 96} & \rightarrow & 4 & \rightarrow & \underline{-\ 4} & \text{or} & \underline{-\ 12} & \rightarrow & \underline{\times\ 4} & \rightarrow & 8{,}448 \\ & & & & 84 & & 84 & & 48 & & \end{array}$$

Time Out: This trick will also work nicely when multiplying numbers that are just under 1,000, except that you must insert a zero into the hundreds place. Let's try it!

Brain Builder

989 × 994

Step 1. Determine the distance 989 is from 1,000: 11.

Step 2. Determine the distance 994 is from 1,000: 6.

Step 3. Subtract 11 from 994 or 6 from 989: $994 - 11$ (or $989 - 6$) $= 983$ (left half of answer).

Step 4. Multiply together the two distances: $11 \times 6 = 066$ (right half of answer).

Step 5. Combine the results of Steps 3 and 4: 983,066 (answer).

Thought Process Summary

$$\begin{array}{rcrcrcrcrcl} & & \text{Distance} & & & & & & & & \\ 989 & \rightarrow & 11 & & 989 & & 994 & & 11 & & \\ \underline{\times\ 994} & \rightarrow & 6 & \rightarrow & \underline{-\ \ 6} & \text{or} & \underline{-\ 11} & \rightarrow & \underline{\times\ 6} & \rightarrow & 983{,}066 \\ & & & & 983 & & 983 & & 066 & & \end{array}$$

Food for Thought: Trick 39 will work as long as the distance multiplication (e.g., 12×4 in Elementary Example 2) does not exceed 99; the related limitation when the multiplied numbers are just under 1,000 (as in our Brain Builder example) is 999. When those limitations are exceeded, the calculation becomes very difficult to execute because carrying is then necessary.

Practice Problems

Begin these exercises by jotting down each distance from 100.

Elementary Exercises

1. 99×93
   9,207
   Combine (99 − 7) and (1 × 7)

2. 92×96

3. 97×94

4. 95×88

5. 95×94

6. 97×89

7. 91×92

8. 94×99

9. 87×97

10. 90×96

11. 93×98

12. 95×86

13. 99×99

14. 85×94

Begin these exercises by jotting down each distance from 1,000.

Brain Builders

1. 994×997

2. 989×993

3. 998×992

4. $\begin{array}{r} 991 \\ \times\ 995 \\ \hline \end{array}$

5. $\begin{array}{r} 996 \\ \times\ 987 \\ \hline \end{array}$

6. $\begin{array}{r} 997 \\ \times\ 996 \\ \hline \end{array}$

7. $\begin{array}{r} 985 \\ \times\ 995 \\ \hline \end{array}$

8. $\begin{array}{r} 994 \\ \times\ 988 \\ \hline \end{array}$

(See solutions on page 211.)

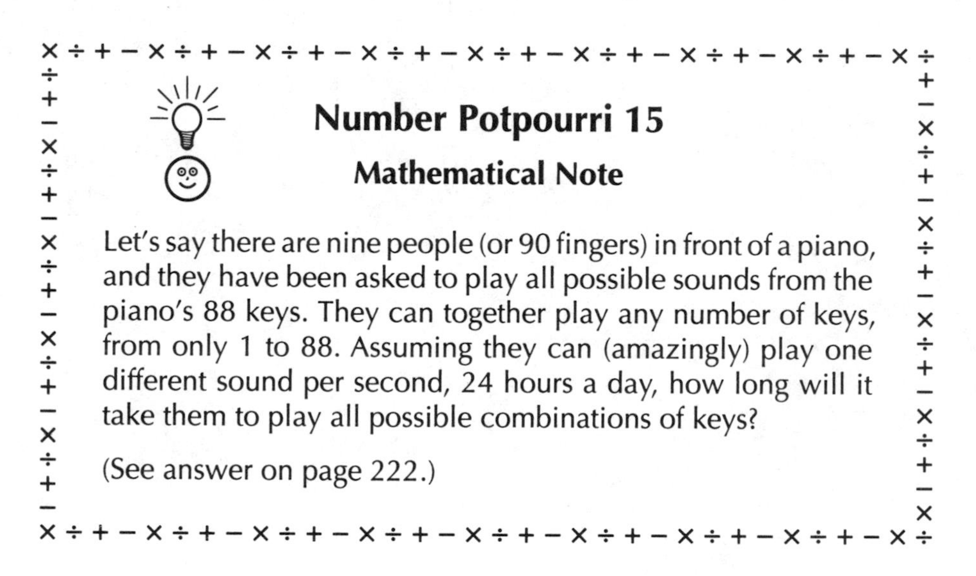

Number Potpourri 15

Mathematical Note

Let's say there are nine people (or 90 fingers) in front of a piano, and they have been asked to play all possible sounds from the piano's 88 keys. They can together play any number of keys, from only 1 to 88. Assuming they can (amazingly) play one different sound per second, 24 hours a day, how long will it take them to play all possible combinations of keys?

(See answer on page 222.)

Parlor Trick 3: The Astounding Cross-Multiplication Trick

Those of you who have read the original *Rapid Math Tricks and Tips* learned how to perform two-digit by two-digit and three-digit by three-digit cross-multiplication. What is so impressive about this technique is that one simply writes the answer without showing any work (all work is done mentally). Four-digit by four-digit cross-multiplication is even more impressive. Let's see how it works.

Demonstration Problem

1,357 x 8,642

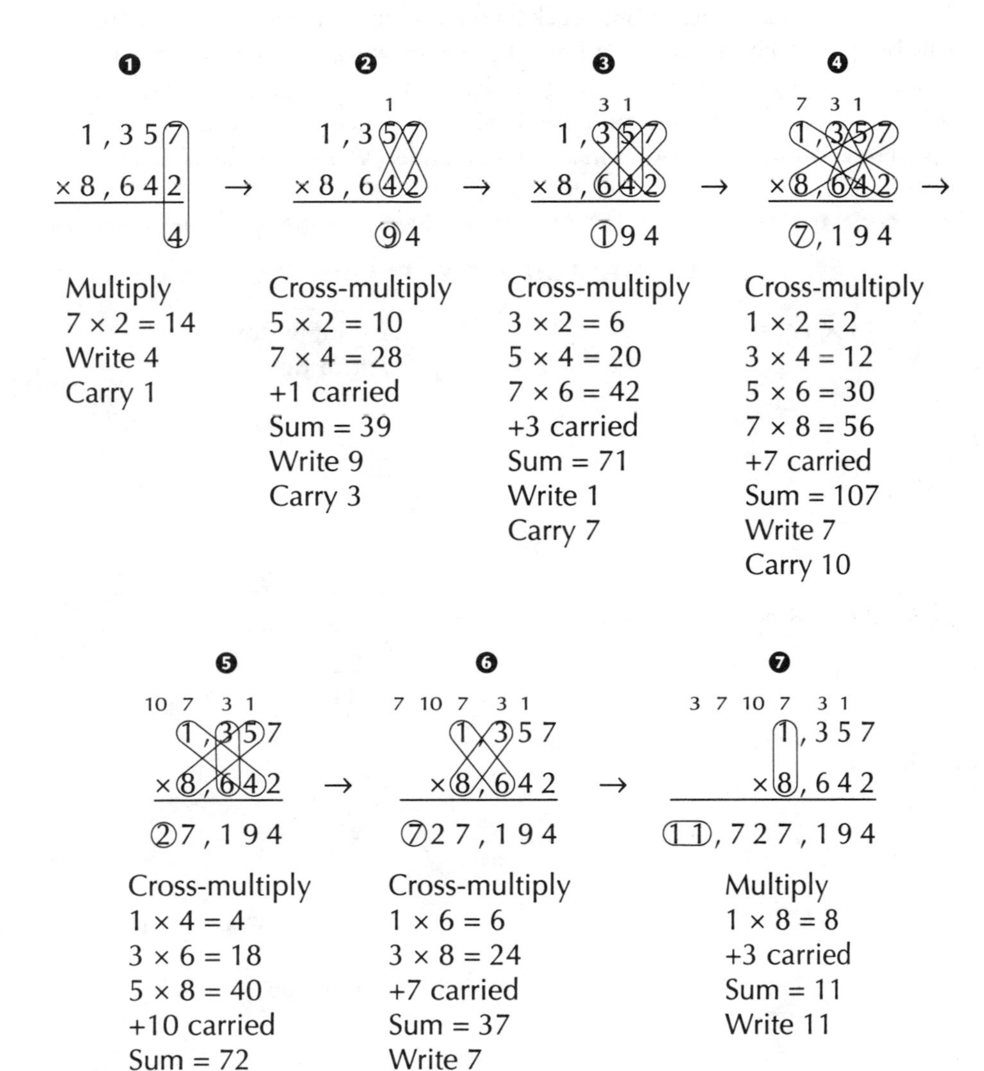

Amaze your family and friends (and yourself) with these:

A.	B.	C.	D.
1,324 × 5,143	8,046 × 3,971	6,273 × 9,506	2,804 × 4,719

(See solutions on page 221.)

Week 3 Quick Quiz

Let's see how many tricks from Week 3 you can remember and apply by taking this brief test. There's no time limit, but try to work through these items as rapidly as possible. Before you begin, glance at the computations and try to identify the trick that you could use. In some instances, however, you will be asked to perform a calculation in a certain manner. When you flip ahead to the solutions, you will see which trick was intended.

Elementary Exercises

1. $13 \times 17 =$

2. $21^2 =$

3. $28 \times 28 =$

4. $33^2 =$

5. $7 \times 33 =$

6. $8 \times 27 =$

7. $26^2 =$

8. $34 \times 34 =$

9. Multiply from left to right:
$$\begin{array}{r} 3{,}142 \\ \times \quad 5 \\ \hline \end{array}$$

10. Do all work in your head:
$$\begin{array}{r} 22 \\ \times\ 54 \\ \hline \end{array}$$

11. $97^2 =$

12. $106 \times 106 =$

13. $$\begin{array}{r} 94 \\ \times\ 97 \\ \hline \end{array}$$

Brain Builders

1. $19 \times 16 =$
2. $72^2 =$
3. $45 \times 22 =$
4. $76^2 =$
5. Multiply from left to right:

$$\begin{array}{r} 7{,}153 \\ \times \quad 6 \\ \hline \end{array}$$

6. $995^2 =$
7. $$\begin{array}{r} 998 \\ \times 993 \\ \hline \end{array}$$

(See solutions on page 219.)

× ÷ + − × ÷ + − × ÷ + − × ÷ + − × ÷ + − × ÷ + − × ÷ + − × ÷ + − × ÷

Number Challenge 3

1. How old do scientists estimate the earth is?

   a. 13 million years c. 4.5 billion years
   b. 750 million years d. 130 billion years

2. How old do scientists estimate the universe is?

   a. 850 million years c. 500 billion years
   b. 14 billion years d. 2.5 trillion years

(See answers on page 222.)

÷ + − × ÷ + − × ÷ + − × ÷ + − × ÷ + − × ÷ + − × ÷ + − × ÷ + − × ÷ +

Week 4 More Unusual Rapid Math Tricks

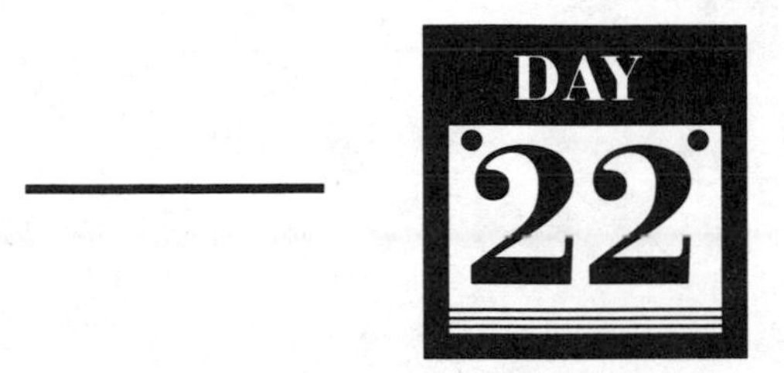

Trick 40: Multiplying Consecutive Numbers Ending in 5

Strategy: This trick involves multiplications such as 15 × 25 and 75 × 65. The secret is to square the number exactly in the middle (which, happily, will always end in a zero). From that product, subtract 25, and you've got your answer! Let's take a look at some examples in a step-by-step manner.

Elementary Example 1

25 × 35

Step 1. Square the number exactly in the middle: $30^2 = 900$.

Step 2. Subtract 25 from the 900: 900 − 25 = 875 (answer).

Thought Process Summary

$$\begin{array}{r} 25 \\ \times\ 35 \\ \hline \end{array} \rightarrow \begin{array}{r} 30 \\ \times\ 30 \\ \hline 900 \end{array} \rightarrow \begin{array}{r} 900 \\ -\ 25 \\ \hline 875 \end{array}$$

Elementary Example 2

65 × 55

Step 1. Square the number exactly in the middle: $60^2 = 3{,}600$.

Step 2. Subtract 25 from the 3,600: 3,600 − 25 = 3,575 (answer).

Thought Process Summary

$$\begin{array}{r} 65 \\ \times\ 55 \\ \hline \end{array} \rightarrow \begin{array}{r} 60 \\ \times\ 60 \\ \hline 3{,}600 \end{array} \rightarrow \begin{array}{r} 3{,}600 \\ -\ \ 25 \\ \hline 3{,}575 \end{array}$$

Brain Builder

105 × 95

Step 1. Square the number exactly in the middle: $100^2 = 10{,}000$.

Step 2. Subtract 25 from the 10,000: $10{,}000 - 25 = 9{,}975$ (answer).

Thought Process Summary

$$\begin{array}{r} 105 \\ \times\ 95 \\ \hline \end{array} \rightarrow \begin{array}{r} 100 \\ \times\ 100 \\ \hline 10{,}000 \end{array} \rightarrow \begin{array}{r} 10{,}000 \\ -\ \ 25 \\ \hline 9{,}975 \end{array}$$

Food for Thought: The number to deduct in Step 2 (25) is easy to remember because half of the distance between the two factors (5), squared, equals 25. Trick 41 (our next trick) deals with multiplication of alternate numbers ending in 5.

Practice Problems

To solve these exercises, just square the number in the middle and subtract 25.

Elementary Exercises

1. $25 \times 15 = \mathbf{20^2 - 25 = 375}$
2. $45 \times 55 =$
3. $35 \times 25 =$
4. $35 \times 45 =$
5. $15 \times 5 =$
6. $15 \times 25 =$

7. $55 \times 65 =$
8. $55 \times 45 =$
9. $45 \times 35 =$
10. $65 \times 55 =$
11. $5 \times 15 =$
12. $25 \times 35 =$

Brain Builders

1. $65 \times 75 =$
2. $95 \times 85 =$
3. $85 \times 75 =$
4. $95 \times 105 =$
5. $85 \times 95 =$
6. $75 \times 85 =$
7. $105 \times 95 =$
8. $75 \times 65 =$

(See solutions on page 212.)

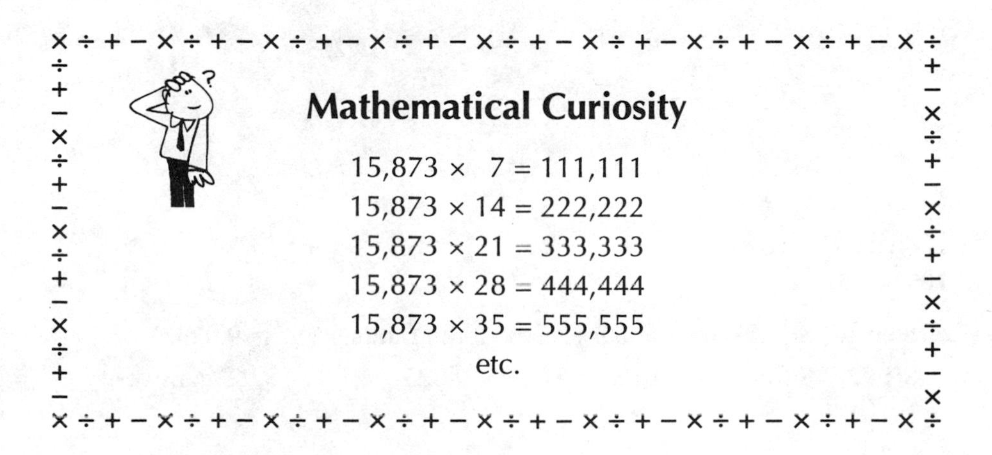

Trick 41: Multiplying Alternate Numbers Ending in 5

Strategy: This trick involves multiplications such as 25 × 45 and 75 × 55. The secret is to square the number exactly in the middle (which will always end in 5). From that product, subtract 100, and you've got your answer! We recommend that you first review Appendix A, page 193 ("Rapidly Square Any Number Ending in 5"), unless you are already comfortable with that trick. As you'll see below, this trick works in a manner similar to Trick 40.

Elementary Example 1

15 × 35

Step 1. Square the number exactly in the middle: $25^2 = 625$.

Step 2. Subtract 100 from the 625: $625 - 100 = 525$ (answer).

Thought Process Summary

$$\begin{array}{r} 15 \\ \times\, 35 \\ \hline \end{array} \rightarrow \begin{array}{r} 25 \\ \times\, 25 \\ \hline 625 \end{array} \rightarrow \begin{array}{r} 625 \\ -\, 100 \\ \hline 525 \end{array}$$

Elementary Example 2

45 × 25

Step 1. Square the number exactly in the middle: $35^2 = 1{,}225$.

Step 2. Subtract 100 from the 1,225: $1{,}225 - 100 = 1{,}125$ (answer).

Thought Process Summary

$$\begin{array}{r} 45 \\ \times\, 25 \\ \hline \end{array} \rightarrow \begin{array}{r} 35 \\ \times\, 35 \\ \hline 1{,}225 \end{array} \rightarrow \begin{array}{r} 1{,}225 \\ -\, 100 \\ \hline 1{,}125 \end{array}$$

Brain Builder

105 × 85

Step 1. Square the number exactly in the middle: $95^2 = 9{,}025$.

Step 2. Subtract 100 from the 9,025: $9{,}025 - 100 = 8{,}925$ (answer).

Thought Process Summary

$$\begin{array}{r} 105 \\ \times\ 85 \\ \hline \end{array} \rightarrow \begin{array}{r} 95 \\ \times\ 95 \\ \hline 9{,}025 \end{array} \rightarrow \begin{array}{r} 9{,}025 \\ -\ 100 \\ \hline 8{,}925 \end{array}$$

Food for Thought: The number to deduct in Step 2 (100) is easy to remember because half of the distance between the two factors (10), squared, equals 100.

Practice Problems

Square the number in the middle and subtract 100 to solve these problems.

Elementary Exercises

1. $25 \times 5 = \mathbf{15^2 - 100 = 125}$
2. $35 \times 55 =$
3. $25 \times 45 =$
4. $45 \times 65 =$
5. $35 \times 15 =$
6. $45 \times 25 =$
7. $5 \times 25 =$
8. $15 \times 35 =$
9. $55 \times 35 =$
10. $65 \times 45 =$

Brain Builders

1. $55 \times 75 =$
2. $85 \times 65 =$
3. $75 \times 95 =$
4. $85 \times 105 =$
5. $95 \times 75 =$
6. $75 \times 55 =$
7. $105 \times 85 =$
8. $85 \times 65 =$

(See solutions on page 212.)

Trick 42: Multiplying Three-Digit Numbers with 0 in the Middle

Strategy: This trick will work for calculations such as 306 × 504 and 701 × 209. One nice feature, as you're about to see, is that you can write the answer from left to right. First, multiply the hundreds digits together, and write the answer in the left-hand portion of the answer space. Then cross-multiply each hundreds digit by the ones digit of the other number. Add the products, and write the sum in the middle of the answer space. Finally, multiply the ones digits together and place that product in the right-hand portion of the answer space. With the exception of the hundreds product in Step 1, you'll need to place a zero to the left of any one-digit amount written in the answer space (so 3, for example, would be written as 03). Let's look at some examples of this very interesting trick.

Elementary Example 1

208 × 507

Step 1. Multiply the hundreds digits together: 2 × 5 = 10.

Step 2. Cross-multiply the hundreds digits by the ones digits: 2 × 7 = 14 and 8 × 5 = 40.

Step 3. Add the products from Step 2: 14 + 40 = 54.

Step 4. Multiply the ones digits together: 8 × 7 = 56.

Step 5. Combine the amounts from Steps 1, 3, and 4: 105,456 (answer).

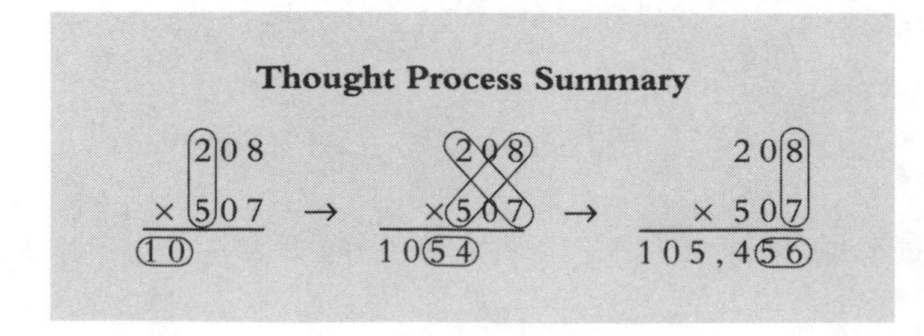

Elementary Example 2

701 × 701

Step 1. Multiply the hundreds digits together: $7 \times 7 = 49$.

Step 2. Cross-multiply the hundreds digits by the ones digits: $7 \times 1 = 7$ and $1 \times 7 = 7$.

Step 3. Add the products from Step 2: $7 + 7 = 14$.

Step 4. Multiply the ones digits together: $1 \times 1 = 01$.

Step 5. Combine the amounts from Steps 1, 3, and 4: 491,401 (answer).

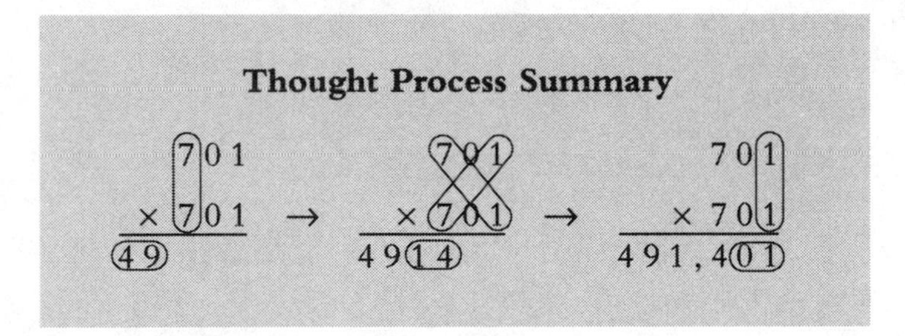

Brain Builder

908 × 605

Step 1. Multiply the hundreds digits together: $9 \times 6 = 54$.

Step 2. Cross-multiply the hundreds digits by the ones digits: $9 \times 5 = 45$ and $8 \times 6 = 48$.

Step 3. Add the products from Step 2: $45 + 48 = 93$.

Step 4. Multiply the ones digits together: $8 \times 5 = 40$.

Step 5. Combine the amounts from Steps 1, 3, and 4: 549,340 (answer).

Thought Process Summary

908 × 605 = 54 → 908 × 605 = 5493 → 908 × 605 = 549,340

Food for Thought: This trick will work as above unless the sum of the cross-multiplications (Step 3) exceeds 99. In the calculation 708×905, for example, the cross-multiplication sum is 107 (7×5 plus 8×9). In such instances, you would have to carry to the ten thousands place, which may convince you to use this trick only when the upper limit of 99 is not exceeded.

Practice Problems

Multiply, cross-multiply, add, and multiply to solve these problems.

Elementary Exercises

1. $\begin{array}{r} 305 \\ \times\ 204 \\ \hline \end{array}$
   62,220
   Combine (3 × 2), (3 × 4) + (5 × 2), and (5 × 4)

2. $\begin{array}{r} 106 \\ \times\ 504 \\ \hline \end{array}$

3. $\begin{array}{r} 403 \\ \times\ 602 \\ \hline \end{array}$

4. $\begin{array}{r} 207 \\ \times\ 205 \\ \hline \end{array}$

5. $\begin{array}{r} 604 \\ \times\ 503 \\ \hline \end{array}$

6. $\begin{array}{r} 803 \\ \times\ 502 \\ \hline \end{array}$

7. $\begin{array}{r} 601 \\ \times\ 305 \\ \hline \end{array}$

8. $\begin{array}{r} 303 \\ \times\ 404 \\ \hline \end{array}$

9. $\begin{array}{r} 507 \\ \times\ 203 \\ \hline \end{array}$

10. $\begin{array}{r} 406 \\ \times\ 306 \\ \hline \end{array}$

11. $\begin{array}{r} 109 \\ \times\ 901 \\ \hline \end{array}$

12. $\begin{array}{r} 802 \\ \times\ 304 \\ \hline \end{array}$

13. $\begin{array}{r} 206 \\ \times\ 403 \\ \hline \end{array}$

14. $\begin{array}{r} 501 \\ \times\ 107 \\ \hline \end{array}$

Brain Builders

1. $\begin{array}{r} 907 \\ \times\ 604 \\ \hline \end{array}$

2. $\begin{array}{r} 603 \\ \times\ 808 \\ \hline \end{array}$

3. $\begin{array}{r} 707 \\ \times\ 508 \\ \hline \end{array}$

4. $\begin{array}{r} 406 \\ \times\ 907 \\ \hline \end{array}$

5. $\begin{array}{r} 309 \\ \times\ 809 \\ \hline \end{array}$

6. $\begin{array}{r} 506 \\ \times\ 408 \\ \hline \end{array}$

7. $\begin{array}{r} 607 \\ \times\ 706 \\ \hline \end{array}$

8. $\begin{array}{r} 909 \\ \times\ 803 \\ \hline \end{array}$

(See solutions on page 212.)

Trick 43: Multiplying with One Number Just under a Whole

Strategy: This trick will work with calculations such as $8 \times 4\frac{3}{4}$ and $12 \times 2\frac{2}{3}$. The secret is to round up the mixed number to the next whole number, multiply, and then subtract the overstated portion. Let's look at some step-by-step explanations below.

Elementary Example 1

$15 \times 3\frac{2}{3}$

Step 1. Round up the mixed number, and multiply: $15 \times 4 = 60$.

Step 2. Multiply 15 by the "added" fraction: $15 \times \frac{1}{3} = 5$.

Step 3. Subtract the 5 from the 60: $60 - 5 = 55$ (answer).

Time Out: What we have done in Elementary Example 1 is to convert the calculation to $15 \times (4 - \frac{1}{3})$, which factors out to $(15 \times 4) - (15 \times \frac{1}{3})$. This works out to $60 - 5$, or 55.

Thought Process Summary

$$\begin{array}{r} 15 \\ \times\ 3\frac{2}{3} \\ \hline \end{array} \rightarrow \begin{array}{r} 15 \\ \times\ 4 \\ \hline 60 \end{array} \rightarrow \begin{array}{r} 15 \\ \times\ \frac{1}{3} \\ \hline 5 \end{array} \rightarrow \begin{array}{r} 60 \\ -\ 5 \\ \hline 55 \end{array}$$

Elementary Example 2

$5\frac{3}{4} \times 8$

Step 1. Round up the mixed number, and multiply: $6 \times 8 = 48$.

Step 2. Multiply 8 by the "added" fraction: $8 \times \frac{1}{4} = 2$.

Step 3. Subtract the 2 from the 48: $48 - 2 = 46$ (answer).

Thought Process Summary

$$\begin{array}{r} 5\frac{3}{4} \\ \times\ 8 \\ \hline \end{array} \rightarrow \begin{array}{r} 6 \\ \times\ 8 \\ \hline 48 \end{array} \rightarrow \begin{array}{r} \frac{1}{4} \\ \times 8 \\ \hline 2 \end{array} \rightarrow \begin{array}{r} 48 \\ -\ 2 \\ \hline 46 \end{array}$$

Brain Builder

$18 \times 2\frac{5}{6}$

Step 1. Round up the mixed number, and multiply: $18 \times 3 = 54$.
Step 2. Multiply 18 by the "added" fraction: $18 \times \frac{1}{6} = 3$.
Step 3. Subtract the 3 from the 54: $54 - 3 = 51$ (answer).

Thought Process Summary

$$\begin{array}{r} 18 \\ \times\, 2\frac{5}{6} \\ \hline \end{array} \rightarrow \begin{array}{r} 18 \\ \times\, 3 \\ \hline 54 \end{array} \rightarrow \begin{array}{r} 18 \\ \times\, \frac{1}{6} \\ \hline 3 \end{array} \rightarrow \begin{array}{r} 54 \\ -\, 3 \\ \hline 51 \end{array}$$

Food for Thought: This basic strategy can also be applied when one number is just over a whole number. For example, $4 \times 3\frac{1}{8}$ can be solved as $(4 \times 3) + (4 \times \frac{1}{8})$, which equals $12\frac{1}{2}$. Similarly, Elementary Example 1 could be solved as $(15 \times 3) + (15 \times \frac{2}{3})$ instead of rounding up and backtracking. Choose whichever technique is more to your liking.

Practice Problems

The operative terms are *round up, multiply,* and *subtract.*

Elementary Exercises

1. $18 \times 2\frac{2}{3} =$
   $\mathbf{(18 \times 3) - (18 \times \frac{1}{3}) = 48}$
2. $5\frac{3}{4} \times 12 =$
3. $25 \times 4\frac{4}{5} =$
4. $3\frac{3}{4} \times 160 =$
5. $90 \times 8\frac{2}{3} =$
6. $5\frac{4}{5} \times 15 =$
7. $14 \times 4\frac{1}{2} =$
8. $6\frac{3}{4} \times 80 =$
9. $5 \times 10\frac{3}{5} =$
10. $4\frac{2}{3} \times 120 =$

11. $4 \times 14\frac{3}{4} =$

12. $9\frac{1}{2} \times 60 =$

13. $30 \times 2\frac{4}{5} =$

14. $4\frac{2}{3} \times 150 =$

Brain Builders

1. $12 \times 4\frac{5}{6} =$

2. $2\frac{7}{8} \times 24 =$

3. $14 \times 3\frac{5}{7} =$

4. $1\frac{8}{9} \times 270 =$

5. $16 \times 4\frac{7}{8} =$

6. $1\frac{5}{6} \times 180 =$

7. $22 \times 3\frac{9}{11} =$

8. $4\frac{7}{8} \times 24 =$

(See solutions on page 213.)

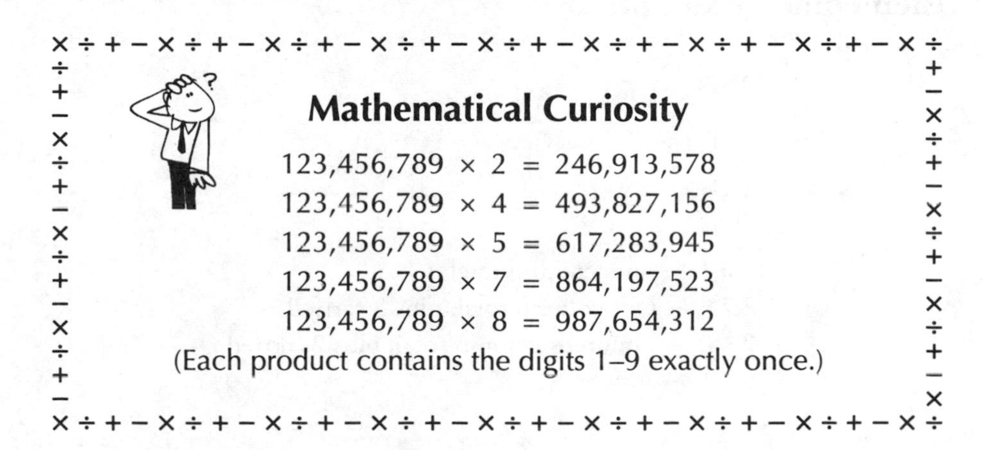

Trick 44: Adding from an Unusual Angle

Strategy: This trick gives the illusion that you are adding without carrying. Actually, the digit you are carrying appears at the bottom of the column rather than at the top. As you will see below, a new digit of the answer appears as each column is added. Let's see how this interesting method of adding works.

Elementary Example 1

```
   926
   591
   804
   177
 + 653
 -----
    21 ← ones column total
   25  ← tens column total, plus 2 carried
 31    ← hundreds column total, plus 2 carried
 -----
 3,151 ← answer
```

(*Note:* Digits in bold constitute the answer.)

Elementary Example 2

```
   288
   645
   109
   337
 + 592
 -----
    31 ← ones column total
   27  ← tens column total, plus 3 carried
 19    ← hundreds column total, plus 2 carried
 -----
 1,971 ← answer
```

Brain Builder

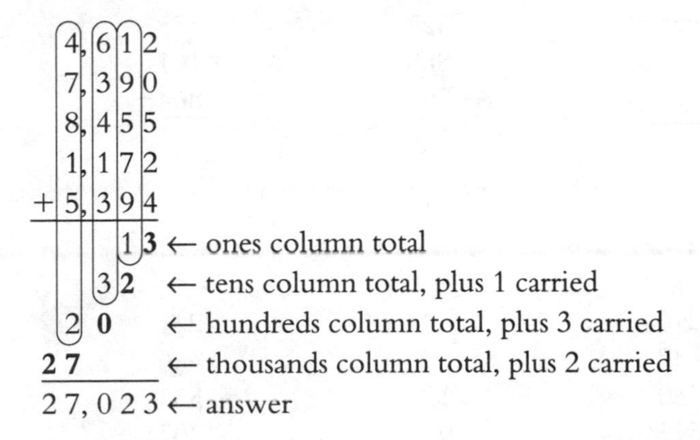

Food for Thought: At first, this trick may appear a bit confusing. But with some practice, you'll find that it will grow on you.

Practice Problems

Don't forget to include the digits placed "below the line" when performing these additions.

Elementary Exercises

1.
```
    327
    145
    809
    633
  + 451
  -----
     25
    16
  23
  -----
  2,365
```

2.
```
    535
    207
    846
    315
  + 494
```

3.
```
    191
    746
    588
    322
  + 407
```

4.
```
    607
    288
    934
    555
  + 327
```

5.
```
    890
    411
    546
    323
  + 147
```

6.
```
    426
    317
    800
    521
  + 799
```

7.
```
    123
    598
    446
    750
  + 316
```

8.
```
    204
    597
    633
    112
  + 824
```

9.	11.	13.
671	139	924
803	746	363
455	527	755
174	888	101
+ 522	+ 361	+ 864

10.	12.	14.
216	604	211
437	778	986
980	123	533
525	955	707
+ 767	+ 846	+ 446

Brain Builders

1.	3.	5.	7.
6,127	3,926	4,160	8,209
3,405	4,134	7,926	7,464
8,993	6,713	3,215	5,315
5,215	5,030	8,347	1,066
+ 7,036	+ 8,872	+ 9,069	+ 7,238

2.	4.	6.	8.
4,257	5,213	1,815	6,467
8,900	8,446	6,709	1,204
3,644	7,207	4,627	3,559
5,113	1,999	8,319	9,311
+ 2,897	+ 6,423	+ 2,045	+ 4,885

(See solutions on page 213.)

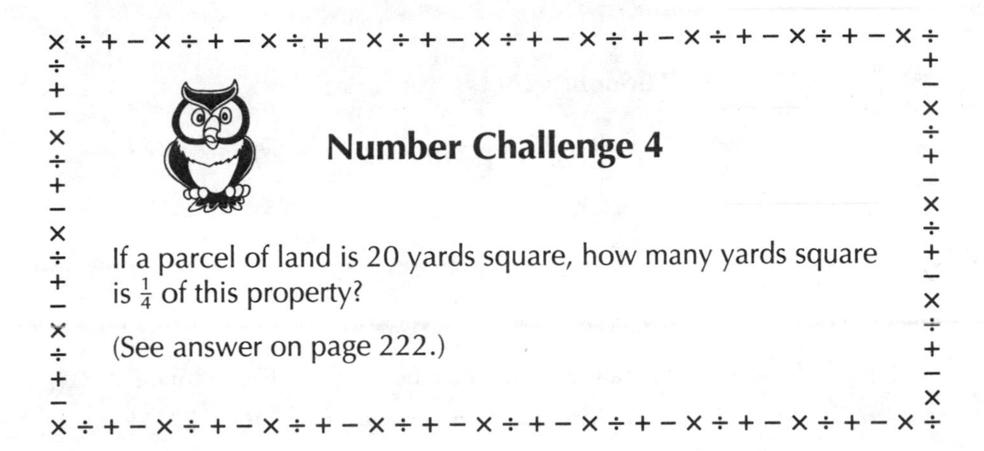

Number Challenge 4

If a parcel of land is 20 yards square, how many yards square is $\frac{1}{4}$ of this property?

(See answer on page 222.)

Trick 45: Adding by Breaking Apart a Number

Strategy: This trick relies upon your ability to determine what number needs to be added to a given number to total 100 (or a multiple of 100). When adding 85 to 137, for example, we first determine that 15 must be added to 85 to total 100. We then split the 137 into (122 + 15), and solve as 122 + 15 + 85, which creates a simple 122 + 100. Read on for some step-by-step illustrations.

Elementary Example 1

137 + 85

Step 1. Determine the number that must be added to 85 to total 100: 15.

Step 2. Break 137 apart, with 15 as one component: 137 = 122 + 15.

Step 3. Restate the calculation: 122 + 15 + 85.

Step 4. Solve: 122 + (15 + 85) = 122 + 100 = 222 (answer).

Thought Process Summary

137 + 85 → (122 + 15) + 85 → 122 + (15 + 85) → 222

Elementary Example 2

91 + 76

Step 1. Determine the number that must be added to 91 to total 100: 9.

Step 2. Break 76 apart, with 9 as one component: 76 = 9 + 67.

Step 3. Restate the calculation: 91 + 9 + 67.

Step 4. Solve: (91 + 9) + 67 = 100 + 67 = 167 (answer).

Thought Process Summary

$91 + 76 \rightarrow 91 + (9 + 67) \rightarrow (91 + 9) + 67 \rightarrow 167$

Brain Builder

246 + 177

Step 1. Determine the number that must be added to 177 to total 200: 23.

Step 2. Break 246 apart, with 23 as one component: 246 = 223 + 23.

Step 3. Restate the calculation: 223 + 23 + 177.

Step 4. Solve: 223 + (23 + 177) = 223 + 200 = 423 (answer).

Thought Process Summary

$246 + 177 \rightarrow (223 + 23) + 177 \rightarrow 223 + (23 + 177) \rightarrow 423$

Food for Thought: In Elementary Example 1, we could have split the 85 into 63 + 22 and then added as follows: (137 + 63) + 22 = 200 + 22 = 222. In other words, when applying this trick, you may break apart whichever number is easier for you.

Practice Problems

Break apart one number or the other to solve these exercises.

Elementary Exercises

1. 175 + 38
   = 175 + (25 + 13) = 213
2. 87 + 46 =
3. 69 + 38 =
4. 188 + 53 =
5. 92 + 79 =
6. 155 + 68 =
7. 89 + 66 =
8. 77 + 53 =

9. 194 + 78 =

10. 84 + 57 =

11. 169 + 62 =

12. 75 + 139 =

13. 67 + 59 =

14. 186 + 48 =

Brain Builders

1. 379 + 143 =

2. 265 + 86 =

3. 146 + 288 =

4. 475 + 257 =

5. 166 + 535 =

6. 339 + 75 =

7. 687 + 233 =

8. 235 + 569 =

(See solutions on page 213.)

Number Potpourri 16

The Mysterious Number 9

When dividing a number by 9, there's a really easy way to figure out (before doing the calculation) what the remainder will be. All you have to do is add the digits of the number you're dividing by 9. For example, the remainder when taking 23 ÷ 9 is 5, because 2 + 3 = 5 (so 23 ÷ 9 = 2 r5). Sometimes the digits of the number total more than 9. When this happens, just add one more time. For example, when taking 67 ÷ 9, the remainder will be 4 because 6 + 7 = 13, and 1 + 3 = 4 (so 67 ÷ 9 = 7 r4). Sometimes the digits of the number total 9 or a multiple of 9 (such as 18 or 27). In these cases, there is no remainder.

Trick 46: Multiplying Two Numbers with a Particular Relationship I

Strategy: This trick will work when you are multiplying two two-digit numbers that possess the following characteristics: One multiplier must have the same tens and ones digit (numbers such as 11, 44, and 77); the other multiplier must contain tens and ones digits that add to 10 (numbers such as 64, 28, and 55). The secret is to pretend that the number whose digits add to 10 is really 10 greater than it is, so, for example, you would pretend that 64 is really 74. Then multiply tens digit by tens digit for the left-hand portion of the answer, and ones digit by ones digit for the right-hand portion. For the latter, a one-digit product is written as two digits. So 4 and 9 would be written as 04 and 09, respectively. Let's apply this interesting trick to some examples.

Elementary Example 1

22 × 73

Step 1. Pretend that the 73 is really 83, and restate the problem: 22 × 83.

Step 2. Multiply the tens digits: 2 × 8 = 16 (left portion of answer).

Step 3. Multiply the ones digits: 2 × 3 = 06 (right portion of answer). The answer is 1,606.

Thought Process Summary

$$\begin{array}{r} 22 \\ \times\, 73 \\ \hline \end{array} \rightarrow \begin{array}{r} 22 \\ \times\, 83 \\ \hline \end{array} \rightarrow \begin{array}{r} 22 \\ \times\, 83 \\ \hline \mathbf{16} \end{array} \rightarrow \begin{array}{r} 22 \\ \times\, 83 \\ \hline 1{,}606 \end{array}$$

Elementary Example 2

44 × 19

Step 1. Pretend that the 19 is really 29, and restate the problem: 44 × 29.

Step 2. Multiply the tens digits: 4 × 2 = 8 (left portion of answer).

Step 3. Multiply the ones digits: 4 × 9 = 36 (right portion of answer). The answer is 836.

Thought Process Summary

$$\begin{array}{r} 44 \\ \times\ 19 \\ \hline \end{array} \rightarrow \begin{array}{r} 44 \\ \times\ 29 \\ \hline \end{array} \rightarrow \begin{array}{r} \mathbf{44} \\ \mathbf{\times\ 29} \\ \hline \mathbf{8} \end{array} \rightarrow \begin{array}{r} \mathbf{44} \\ \mathbf{\times\ 29} \\ \hline \mathbf{836} \end{array}$$

Brain Builder

99 × 64

Step 1. Pretend that the 64 is really 74, and restate the problem: 99 × 74.

Step 2. Multiply the tens digits: 9 × 7 = 63 (left portion of answer).

Step 3. Multiply the ones digits: 9 × 4 = 36 (right portion of answer). The answer is 6,336.

Thought Process Summary

$$\begin{array}{r} 99 \\ \times\ 64 \\ \hline \end{array} \rightarrow \begin{array}{r} 99 \\ \times\ 74 \\ \hline \end{array} \rightarrow \begin{array}{r} \mathbf{99} \\ \mathbf{\times\ 74} \\ \hline \mathbf{63} \end{array} \rightarrow \begin{array}{r} \mathbf{99} \\ \mathbf{\times\ 74} \\ \hline \mathbf{6{,}336} \end{array}$$

Food for Thought: You, of course, could have multiplied the ones digits before multiplying the tens digits (i.e., Steps 2 and 3 could be reversed). However, multiplying the tens digits first enables you to write the answer, or to say the answer aloud, from left to right.

Practice Problems

First augment by 10, then multiply twice, and you're all set!

Elementary Exercises

1. $33 \times 28 \rightarrow \mathbf{33 \times 38 \rightarrow 924}$
2. $55 \times 46 =$
3. $11 \times 37 =$
4. $44 \times 91 =$
5. $22 \times 55 =$
6. $66 \times 82 =$
7. $33 \times 64 =$
8. $55 \times 19 =$
9. $22 \times 73 =$
10. $11 \times 28 =$
11. $66 \times 55 =$
12. $44 \times 46 =$
13. $33 \times 37 =$
14. $55 \times 55 =$

Brain Builders

1. $88 \times 91 =$
2. $77 \times 28 =$
3. $99 \times 46 =$
4. $66 \times 73 =$
5. $77 \times 19 =$
6. $99 \times 55 =$
7. $88 \times 82 =$
8. $77 \times 37=$

(See solutions on page 214.)

Trick 47: Multiplying Two Numbers with a Particular Relationship II

Strategy: This trick applies when you are multiplying two two-digit numbers whose tens digits add to 10 and whose ones digits are identical. Calculations such as 32×72 and 46×66 qualify. The secret is to first multiply the tens digits and then add the ones digit. This will produce the left half of the answer. Then, square the ones digit for the answer's right half. For 1^2, 2^2, and 3^2, write 01, 04, and 09, respectively. Let's take a look at some examples.

Elementary Example 1

63 × 43

Step 1. Multiply the tens digits: $6 \times 4 = 24$.

Step 2. Add the ones digit (3) to the 24: $24 + 3 = 27$.

Step 3. Square the ones digit: $3 \times 3 = 09$.

Step 4. Combine the amounts from Steps 2 and 3: 2,709 (answer).

Thought Process Summary

$$\begin{array}{r}63\\ \times\,43\\ \hline \end{array} \rightarrow \begin{array}{r}6\\ \times\,4\\ \hline 24\end{array} \rightarrow \begin{array}{r}24\\ +\,3\\ \hline 27\end{array} \rightarrow \begin{array}{r}3\\ \times\,3\\ \hline 09\end{array} \rightarrow 2{,}709$$

Elementary Example 2

35 × 75

Step 1. Multiply the tens digits: $3 \times 7 = 21$.

Step 2. Add the ones digit (5) to the 21: $21 + 5 = 26$.

Step 3. Square the ones digit: $5 \times 5 = 25$.

Step 4. Combine the amounts from Steps 2 and 3: 2,625 (answer).

Thought Process Summary

$$\begin{array}{r}35\\ \times\,75\\ \hline \end{array} \rightarrow \begin{array}{r}3\\ \times\,7\\ \hline 21\end{array} \rightarrow \begin{array}{r}21\\ +\,5\\ \hline 26\end{array} \rightarrow \begin{array}{r}5\\ \times\,5\\ \hline 25\end{array} \rightarrow 2{,}625$$

Brain Builder

89 × 29

Step 1. Multiply the tens digits: $8 \times 2 = 16$.
Step 2. Add the ones digit (9) to the 16: $16 + 9 = 25$.
Step 3. Square the ones digit: $9 \times 9 = 81$.
Step 4. Combine the amounts from Steps 2 and 3: 2,581 (answer).

Thought Process Summary

$$\begin{array}{r}89\\ \times 29\\ \hline\end{array} \rightarrow \begin{array}{r}8\\ \times 2\\ \hline 16\end{array} \rightarrow \begin{array}{r}16\\ +\ 9\\ \hline 25\end{array} \rightarrow \begin{array}{r}9\\ \times 9\\ \hline 81\end{array} \rightarrow 2{,}581$$

Food for Thought: The original *Rapid Math Tricks and Tips* covers the reverse of this trick, whereby the sum of the ones digits is 10 and the tens digits are identical.

Practice Problems

The steps to follow are multiply, add, square, and combine.

Elementary Exercises

1. 72 × 32 =
   Combine (7 × 3) + 2 and (2 × 2) = 2,304
2. 65 × 45 =
3. 11 × 91 =
4. 24 × 84 =
5. 53 × 53 =
6. 36 × 76 =
7. 82 × 22 =
8. 41 × 61 =
9. 75 × 35 =
10. 93 × 13 =

11. $54 \times 54 =$

12. $46 \times 66 =$

13. $71 \times 31 =$

14. $85 \times 25 =$

Brain Builders

1. $69 \times 49 =$

2. $28 \times 88 =$

3. $59 \times 59 =$

4. $77 \times 37 =$

5. $97 \times 17 =$

6. $58 \times 58 =$

7. $89 \times 29 =$

8. $47 \times 67 =$

(See solutions on page 214.)

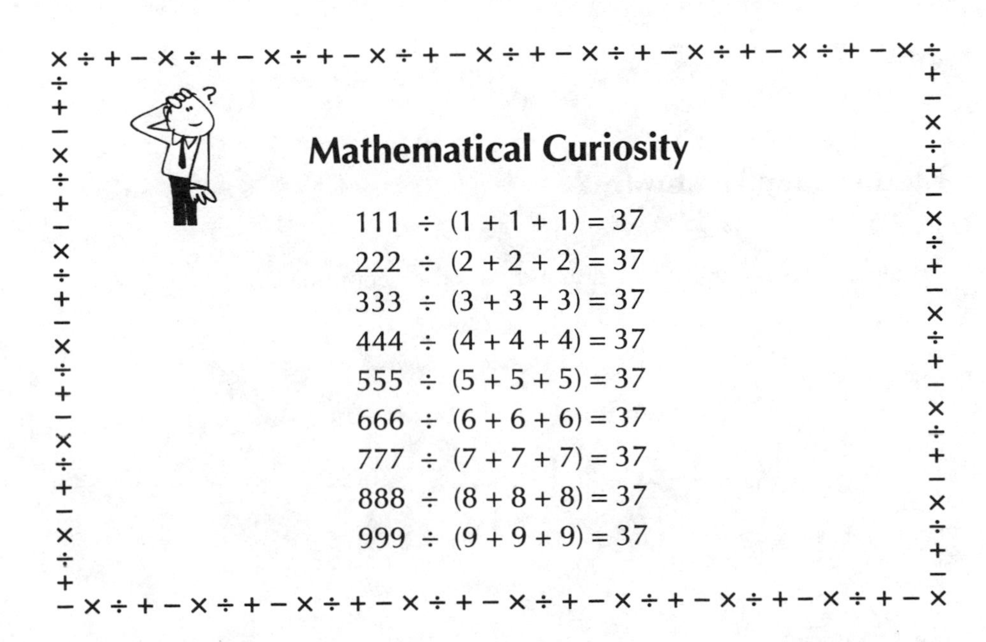

Mathematical Curiosity

$111 \div (1 + 1 + 1) = 37$
$222 \div (2 + 2 + 2) = 37$
$333 \div (3 + 3 + 3) = 37$
$444 \div (4 + 4 + 4) = 37$
$555 \div (5 + 5 + 5) = 37$
$666 \div (6 + 6 + 6) = 37$
$777 \div (7 + 7 + 7) = 37$
$888 \div (8 + 8 + 8) = 37$
$999 \div (9 + 9 + 9) = 37$

Trick 48: Multiplying by Fractions

Strategy: When multiplying a whole number by a fraction, it's okay to move the denominator of the fraction underneath the whole number, and it's desirable if it will simplify the calculation. For example, the calculation $\frac{3}{5} \times 45$ may appear a bit intimidating. However, if you move the denominator under the whole number, you have a much simpler computation at hand: $3 \times \frac{45}{5}$ (or 3×9). Let's look at some more examples of this unusually simple technique.

Elementary Example 1
$\frac{2}{3} \times 36$

Step 1. Recast the calculation, moving the denominator: $2 \times \frac{36}{3}$.

Step 2. Solve: $2 \times \frac{36}{3} = 2 \times 12 = 24$ (answer).

Thought Process Summary

$$\frac{2}{3} \times 36 \rightarrow 2 \times \frac{36}{3} \rightarrow 2 \times 12 \rightarrow 24$$

Elementary Example 2
$52 \times \frac{3}{4}$

Step 1. Recast the calculation, moving the denominator: $\frac{52}{4} \times 3$.

Step 2. Solve: $\frac{52}{4} \times 3 = 13 \times 3 = 39$ (answer).

Thought Process Summary

$$52 \times \frac{3}{4} \rightarrow \frac{52}{4} \times 3 \rightarrow 13 \times 3 \rightarrow 39$$

Brain Builder

$\frac{5}{7} \times 98$

Step 1. Recast the calculation, moving the denominator: $5 \times \frac{98}{7}$.

Step 2. Solve: $5 \times \frac{98}{7} = 5 \times 14 = 70$ (answer).

Thought Process Summary

$$\frac{5}{7} \times 98 \rightarrow 5 \times \frac{98}{7} \rightarrow 5 \times 14 \rightarrow 70$$

Food for Thought: When using this technique, the numbers won't always work out as nicely as in the examples above. However, the restatement may still make life easier. For example, when solving for $\frac{3}{8} \times 20$, you would recast the problem as $3 \times \frac{20}{8}$, which equals $3 \times 2\frac{1}{2}$, or $7\frac{1}{2}$. Not as user-friendly as our other examples, but the trick still works.

Practice Problems

Moving the denominator will simplify these practice problems.

Elementary Exercises

1. $\frac{3}{5} \times 75 =$ **$3 \times \frac{75}{5} = 45$**
2. $\frac{3}{4} \times 36 =$
3. $42 \times \frac{2}{3} =$
4. $60 \times \frac{4}{5} =$
5. $32 \times \frac{3}{4} =$
6. $\frac{2}{3} \times 57 =$
7. $\frac{3}{5} \times 55 =$
8. $72 \times \frac{5}{6} =$
9. $\frac{2}{3} \times 180 =$
10. $\frac{3}{4} \times 140 =$
11. $\frac{4}{5} \times 125 =$
12. $200 \times \frac{3}{5} =$
13. $88 \times \frac{3}{4} =$
14. $\frac{5}{6} \times 180 =$

Brain Builders

1. $56 \times \frac{5}{8} =$
2. $49 \times \frac{6}{7} =$
3. $\frac{2}{3} \times 84 =$
4. $\frac{7}{8} \times 400 =$
5. $\frac{5}{7} \times 280 =$
6. $\frac{3}{4} \times 22 =$
7. $540 \times \frac{7}{9} =$
8. $210 \times \frac{4}{7} =$

(See solutions on page 214.)

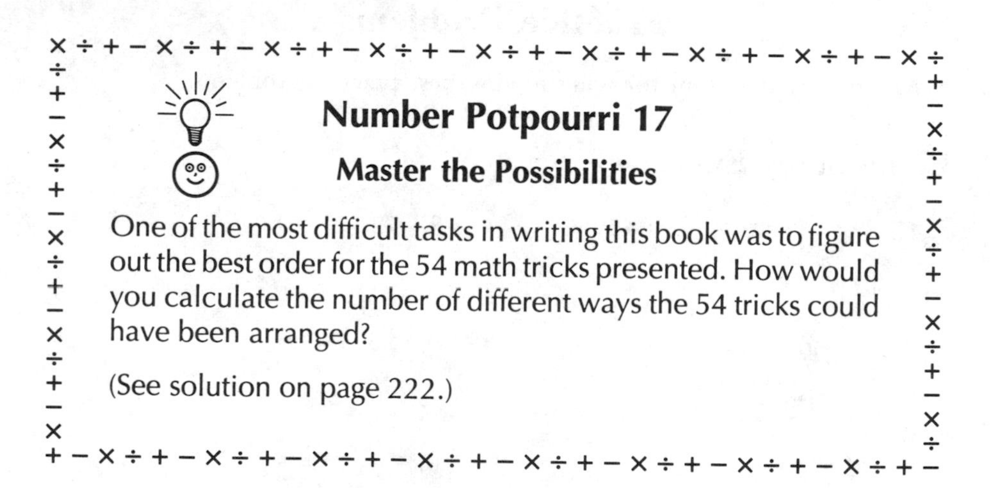

Number Potpourri 17

Master the Possibilities

One of the most difficult tasks in writing this book was to figure out the best order for the 54 math tricks presented. How would you calculate the number of different ways the 54 tricks could have been arranged?

(See solution on page 222.)

Trick 49: Multiplying in Chunks

Strategy: When multiplying, don't hesitate to operate on two or three digits at once, where manageable. Let's consider the calculation 715×4. Ordinarily, you would take $5 \times 4 = 20$ (write 0, carry 2), and so forth. Instead, why not just take $15 \times 4 = 60$ (write 60, carry nothing). Then take $7 \times 4 = 28$, for an answer of 2,860. What about $28{,}125 \times 3$? Just take $125 \times 3 = 375$, and go from there! The trick is to spot "chunks" of numbers within other numbers that can easily be treated as one unit.

Elementary Example 1

812 × 6

Step 1. Multiply 12 by 6: $12 \times 6 = 72$ (right portion of answer).

Step 2. Multiply 8 by 6: $8 \times 6 = 48$ (left portion of answer). The answer is 4,872.

Thought Process Summary

$$\begin{array}{r} 812 \\ \times\ 6 \\ \hline \end{array} \rightarrow \begin{array}{r} 812 \\ \times\ \mathbf{6} \\ \hline \mathbf{72} \end{array} \rightarrow \begin{array}{r} \mathbf{812} \\ \times\ \mathbf{6} \\ \hline \mathbf{4{,}872} \end{array}$$

Elementary Example 2

227 × 4

Step 1. Multiply 7 by 4: $7 \times 4 = 28$ (write 8, carry 2).

Step 2. Multiply 22 by 4, plus 2 carried: $(22 \times 4) + 2 = 90$. The answer is 908.

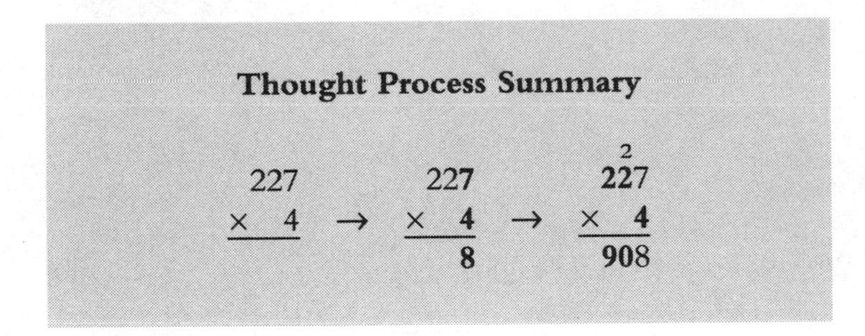

Brain Builder

15,120 × 5

Step 1. Multiply 120 by 5: $120 \times 5 = 600$ (right portion of answer).

Step 2. Multiply 15 by 5: $15 \times 5 = 75$ (left portion of answer). The answer is 75,600.

Thought Process Summary

$$\begin{array}{r} 15{,}120 \\ \times\ 5 \\ \hline \end{array} \rightarrow \begin{array}{r} 15{,}\mathbf{120} \\ \times\ \mathbf{5} \\ \hline \mathbf{600} \end{array} \rightarrow \begin{array}{r} \mathbf{15}{,}120 \\ \times\ \mathbf{5} \\ \hline \mathbf{75}{,}600 \end{array}$$

Food for Thought: Most people can perform "chunk multiplication" on calculations such as 11×8, 25×4, and 32×3. However, for computations such as 17×7, 33×4, and 23×5, it's probably best to operate on one digit at a time.

Practice Problems

Try multiplying two digits (or even three) at a time, wherever possible.

Elementary Exercises

1. 715×5
   3,575
   (15 × 5), then (7 × 5)
2. 511×9
3. 135×4
4. 122×4
5. 921×3
6. 710×8
7. 114×7
8. 523×3
9. 612×6
10. 218×4
11. 813×5
12. 115×3
13. 812×8
14. 715×6

Brain Builders

1. $11{,}150 \times 4$

2. $81{,}212 \times 5$

3. $22{,}321 \times 3$

4. $13{,}120 \times 6$

5. $7{,}233 \times 3$

6. $8{,}250 \times 4$

7. $15{,}125 \times 3$

8. $71{,}211 \times 8$

(See solutions on page 215.)

Number Potpourri 18

Numbers Gone Mad

Here's a strange one. Take any number, any number at all. Subtract the sum of the number's digits. What you'll get, without fail, is an answer that can be divided evenly by 9. For example, take the number 736. The sum of the digits (7 + 3 + 6) is 16. Subtract 16 from 736, and you get 720, which can be divided evenly by 9 (720 ÷ 9 = 80 exactly). Note also that the digit-sum of our number 720 (7 + 2 + 0) is 9!

Trick 50: Multiplying a Two-Digit Number by 111

Strategy: For those of you who read the original *Rapid Math Tricks and Tips*, Trick 50 is very similar to the marvelous "multiply by 11" trick (which, incidentally, is reproduced in Appendix A, page 194, for your perusal). When multiplying a two-digit number by 111, however, the two digits must add to 9 or less. For example, the number 54 works, but not 28. First, write the two-digit number in the answer space, leaving a two-digit space between the two digits. Then, insert the sum of the number's digits twice in-between the two digits themselves. It seems almost too good to be true, as you'll see below.

Elementary Example 1

63 × 111

Step 1. Write the 63 in the answer space, leaving some room between the two digits: 6 $\underline{?}$ $\underline{?}$ 3.

Step 2. Add the two digits of 63: 6 + 3 = 9.

Step 3. Insert the 9 twice in the answer space: 6,993 (answer).

Thought Process Summary

63 × 111 → 6 $\underline{?}$ $\underline{?}$ 3 → 6 + 3 = 9 → 6,993.

Elementary Example 2

25 × 111

Step 1. Write the 25 in the answer space, leaving some room between the two digits: 2 $\underline{?}$ $\underline{?}$ 5.

Step 2. Add the two digits of 25: 2 + 5 = 7.

Step 3. Insert the 7 twice in the answer space: 2,775 (answer).

Thought Process Summary

25 × 111 → 2 ? ? 5 → 2 + 5 = 7 → 2,775

Time Out: Want to multiply a two-digit number by 1,111? Just solve as above, except insert the digit-sum three times.

Brain Builder

44 × 1,111

Step 1. Write the 44 in the answer space, leaving some room between the two digits: 4 ? ? ? 4.

Step 2. Add the two digits of 44: 4 + 4 = 8.

Step 3. Insert the 8 three times in the answer space: 48,884 (answer).

Thought Process Summary

44 × 1,111 → 4 ? ? ? 4 → 4 + 4 = 8 → 48,884

Food for Thought: When the digits of the number to be multiplied by 111 (or 1,111, etc.) total 10 or more, the trick is ruined because multi-carrying then takes place. Actually, this limitation can be overcome by applying Trick 51 (as explained in that trick's "Food for Thought" section).

Practice Problems

Just add the digits and insert the total the appropriate number of times.

Elementary Exercises

1. 35 × 111 =
   Insert (3 + 5) twice = 3,885
2. 54 × 111 =
3. 23 × 111 =
4. 81 × 111 =
5. 111 × 43 =
6. 111 × 72 =

7. $111 \times 62 =$

8. $111 \times 18 =$

9. $26 \times 111 =$

10. $53 \times 111 =$

11. $45 \times 111 =$

12. $61 \times 111 =$

13. $111 \times 15 =$

14. $111 \times 33 =$

Brain Builders

1. $1{,}111 \times 25 =$

2. $1{,}111 \times 36 =$

3. $1{,}111 \times 42 =$

4. $1{,}111 \times 52 =$

5. $63 \times 1{,}111 =$

6. $13 \times 1{,}111 =$

7. $21 \times 1{,}111 =$

8. $32 \times 1{,}111 =$

(See solutions on page 215.)

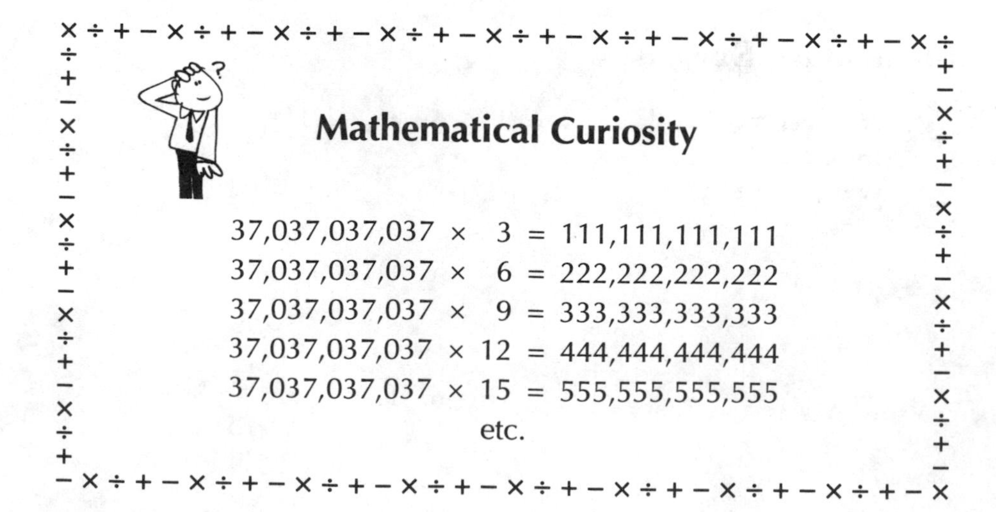

Mathematical Curiosity

$$37{,}037{,}037{,}037 \times 3 = 111{,}111{,}111{,}111$$
$$37{,}037{,}037{,}037 \times 6 = 222{,}222{,}222{,}222$$
$$37{,}037{,}037{,}037 \times 9 = 333{,}333{,}333{,}333$$
$$37{,}037{,}037{,}037 \times 12 = 444{,}444{,}444{,}444$$
$$37{,}037{,}037{,}037 \times 15 = 555{,}555{,}555{,}555$$

etc.

Trick 51: Multiplying Three-Digit or Larger Numbers by 111

Strategy: This is an advanced variation on Trick 50. To multiply a three-digit or larger number by 111, first carry down the ones digit of the number. Then add together the ones and tens digits, and so on, writing down in the answer space only the right-hand digit each time (until the last addition). It will be necessary to carry when any sum exceeds 9. As you'll see below, this impressive trick is a lot easier to perform than it sounds.

Elementary Example 1

234 × 111

Step 1. Write 4 in the answer space as the ones digit.

Step 2. Add: 4 + 3 = 7 (tens-digit answer).

Step 3. Add: 4 + 3 + 2 = 9 (hundreds-digit answer).

Step 4. Add: 3 + 2 = 5 (thousands-digit answer).

Step 5. Write 2 in the answer space as the ten thousands digit.

Step 6. Summarize. The answer is 25,974.

Thought Process Summary

234		234		234		234		234
× 111	→	× 111	→	× 111	→	× 111	→	× 111
4		74		974		5,974		25,974

Elementary Example 2

526 × 111

Step 1. Write 6 in the answer space as the ones digit.

Step 2. Add: 6 + 2 = 8 (tens-digit answer).

Step 3. Add: 6 + 2 + 5 = 13 (3 is the hundreds-digit answer; carry the 1).

Step 4. Add: 2 + 5 + 1 carried = 8 (thousands-digit answer).

Step 5. Write 5 in the answer space as the ten-thousands–digit answer.

Step 6. Summarize. The answer is 58,386.

Thought Process Summary

526 × 111 = 6	→	526 × 111 = 86	→	526 × 111 = 386	→	526 × 111 = 8,386 (carry 1)	→	526 × 111 = 58,386 (carry 1)

Brain Builder

8,075 × 111

Step 1. Write 5 in the answer space as the ones digit.

Step 2. Add: 5 + 7 = 12 (2 is the tens-digit answer; carry the 1).

Step 3. Add: 5 + 7 + 0 + 1 carried = 13 (3 is the hundreds-digit answer; carry the 1).

Step 4. Add: 7 + 0 + 8 + 1 carried = 16 (6 is the thousands-digit answer; carry the 1).

Step 5. Add: 0 + 8 + 1 carried = 9 (ten-thousands–digit answer).

Step 6. Write 8 in the answer space as the hundred-thousands–digit answer.

Step 7. Summarize. The answer is 896,325.

Thought Process Summary

8,075 × 111 = 5	→	8,075 × 111 = 25	→	8,075 × 111 = 325 (carry 1)	→
8,075 × 111 = 6,325 (carries 1 1)	→	8,075 × 111 = 96,325 (carries 1 1 1)	→	8,075 × 111 = 896,325 (carries 1 1 1)	

Food for Thought: You can use Trick 51 to overcome the limitation of Trick 50, whereby the digit-sum of the two-digit factor exceeds 9. For example, if you were multiplying 85 by 111, simply pretend that the 85 is written as 085, and proceed as above. Try it—you'll see that it works just fine!

Practice Problems

Add digits from right to left to work these exercises.

Elementary Exercises

1. 312 × 111 =
   2, 2 + 1, 2 + 1 + 3,
   1 + 3, 3 = 34,632
2. 143 × 111 =
3. 431 × 111 =
4. 342 × 111 =
5. 222 × 111 =
6. 536 × 111 =
7. 704 × 111 =
8. 111 × 365 =
9. 111 × 418 =
10. 111 × 829 =
11. 111 × 663 =
12. 111 × 219 =
13. 111 × 536 =
14. 111 × 777 =

Brain Builders

1. 1,234 × 111 =
2. 4,132 × 111 =
3. 3,207 × 111 =
4. 4,166 × 111 =
5. 111 × 7,460 =
6. 111 × 1,926 =
7. 111 × 3,199 =
8. 111 × 5,988 =

(See solutions on page 215.)

Trick 52: Multiplying by 999

Strategy: Now that you've learned how to rapidly multiply by 111, let's go to the other extreme — multiplication by 999. To multiply a number by 999, first subtract 1 from the number to obtain the left-hand portion of the answer. Then subtract the number from 1,000 to obtain the right-hand portion. That's all there is to it. (*Hint:* It is faster to subtract by adding. For example, to subtract 775 from 1,000, ask yourself, "775 plus what equals 1,000?") Let's take a look at this trick in action.

Elementary Example 1

775 × 999

Step 1. Subtract: 775 − 1 = 774 (left portion of answer).

Step 2. Subtract: 1,000 − 775 = 225 (right portion of answer).

Step 3. Combine: 774,225 is the answer.

Thought Process Summary

$$\begin{array}{r} 775 \\ \times\, 999 \\ \hline \end{array} \rightarrow \begin{array}{r} 775 \\ -\ \ 1 \\ \hline 774 \end{array} \rightarrow \begin{array}{r} 1{,}000 \\ -\ 775 \\ \hline 225 \end{array} \rightarrow 774{,}225$$

Elementary Example 2

88 × 999

Step 1. Subtract: 88 − 1 = 87 (left portion of answer).

Step 2. Subtract: 1,000 − 88 = 912 (right portion of answer).

Step 3. Combine: 87,912 is the answer.

Thought Process Summary

$$\begin{array}{r} 88 \\ \times\, 999 \\ \hline \end{array} \rightarrow \begin{array}{r} 88 \\ -\ 1 \\ \hline 87 \end{array} \rightarrow \begin{array}{r} 1{,}000 \\ -\quad 88 \\ \hline 912 \end{array} \rightarrow 87{,}912$$

Brain Builder

9 × 999

Step 1. Subtract: 9 − 1 = 8 (left portion of answer).

Step 2. Subtract: 1,000 − 9 = 991 (right portion of answer).

Step 3. Combine: 8,991 is the answer.

Thought Process Summary

$$\begin{array}{r} 9 \\ \times\, 999 \\ \hline \end{array} \rightarrow \begin{array}{r} 9 \\ -\ 1 \\ \hline 8 \end{array} \rightarrow \begin{array}{r} 1{,}000 \\ -\quad 9 \\ \hline 991 \end{array} \rightarrow 8{,}991$$

Food for Thought: What you are doing, in effect, is multiplying by (1,000 − 1). All we have done here is to show you a more convenient way to approach the calculation. To master this trick, you'll have to practice subtracting one-, two-, and three-digit numbers from 1,000. But once you get the hang of it, you can astound others with your enormous mental math powers!

Practice Problems

As you begin these exercises, don't accidentally subtract 1 from the 999.

Elementary Exercises

1. 999 × 950 =
   Combine (950 − 1) and (1,000 − 950) = 949,050
2. 999 × 825 =
3. 999 × 730 =
4. 999 × 480 =

5. $926 \times 999 =$
6. $299 \times 999 =$
7. $153 \times 999 =$
8. $608 \times 999 =$
9. $999 \times 90 =$
10. $999 \times 75 =$
11. $999 \times 66 =$
12. $999 \times 85 =$
13. $71 \times 999 =$
14. $89 \times 999 =$

Brain Builders

1. $999 \times 7 =$
2. $999 \times 13 =$
3. $999 \times 27 =$
4. $999 \times 8 =$
5. $36 \times 999 =$
6. $17 \times 999 =$
7. $41 \times 999 =$
8. $6 \times 999 =$

(See solutions on page 216.)

Parlor Trick 4: The Phenomenal Fifth-Root Trick—Expanded Version

In the original *Rapid Math Tricks and Tips,* we presented a trick whereby one extracts the fifth root of a very long number. We will reproduce that trick here but will expand it to more than double its original magnitude.

What is the fifth root of a number? It's much like the square root, only taken a few steps further. For example, $7^5 = 7 \times 7 \times 7 \times 7 \times 7$, which equals

16,807. Therefore, the fifth root of 16,807 is 7. Similarly, $24^5 = 24 \times 24 \times 24 \times 24 \times 24 = 7{,}962{,}624$. Therefore, the fifth root of 7,962,624 is 24.

Ask someone to multiply a whole number (from 1 to 99) by itself five times, as shown above. An 8-digit calculator will be able to accommodate up to 39^5, whereas a 10-digit calculator will be able to accommodate up to 99^5. Since most calculators accommodate only 8 digits, you may wish to purchase one with a 10-digit display to allow up to 99^5. Obviously, don't watch as your volunteer is performing the calculation, but write down the final product when the calculation has been completed. It should take you no longer than a few seconds at that point to extract the fifth root.

Strategy: Let's take a couple of examples. Suppose someone has performed the requisite calculation and obtains a product of 32,768. Within about 3 seconds you will know that the fifth root of that number is 8.

Here's the Trick: First of all, the ones-place digit in the product will automatically become the ones-place digit of the answer. Since 8 is the ones digit of 32,768, the ones digit of the fifth root will also be 8. Next, completely ignore the next four digits to the left of the ones digit (i.e., the tens, hundreds, thousands, and ten-thousands digits). It may be easier to write the product down and cross out the digits with a pencil than to just ignore them because you will then concentrate solely on the digits that remain. In the above case, no digits remain after crossing out the four digits, so the answer is simply 8.

To extract a two-digit fifth root, you'll need to have the following information memorized. If no number remains, then the answer is a one-digit number.

If the remaining number is in the following range:		Then the tens digit is:
1–30	⟶	1
30–230	⟶	2
230–1,000	⟶	3
1,000–3,000	⟶	4
3,000–7,500	⟶	5
7,500–16,000	⟶	6
16,000–32,000	⟶	7
32,000–57,000	⟶	8
57,000–99,000	⟶	9

Now let's take a number with a two-digit fifth root. Suppose someone performs the necessary calculation and arrives at 69,343,957. What will be the ones digit of the fifth root? That's right, it will be 7. Next, ignore or cross out the next four digits (i.e., the 4395). What remains is the number 693.

This number is in the 230–1,000 range; therefore, the tens digit is 3, and the answer is 37. You might be wondering what to do if the remaining number is, for example, 230. Will the tens digit be 2 or 3? You don't have to worry because the remaining number will never be any of the border numbers.

When using the ranges above, it might be easiest to count off each number with your fingers, as follows: 1–30–230–1,000–3,000–7,500–16,000–32,000–57,000, until you reach the range containing the remaining number at hand. For example, we would have counted 1–30–230 in the above example, indicating a tens digit of 3. (We stop at 230 because the next number in the series, 1,000, exceeds the remaining number, 693.)

Try another exercise: Extract the fifth root of 7,339,040,224. You know that the ones digit of the answer is 4. Crossing out the next four digits, the remaining number is 73,390. Referring to the above ranges, you can see that the tens digit is 9, and the answer is 94.

Now it's your turn to extract the fifth root of:

A. 7,776
B. 844,596,301
C. 3,276,800,000
D. 459,165,024
E. 20,511,149
F. 371,293
G. 7,737,809,375
H. 2,887,174,368
I. 130,691,232
J. 79,235,168
K. 16,807
L. 9,509,900,499

(See solutions on page 221.)

Expanded Version

In our original version above, we used a 10-digit calculator, having our subject take a number as high as 99 to the fifth power. In our expanded version, we will have a 12-digit calculator handy (they're not very expensive to purchase) and have our subject take a number between 99 and 252 to the fifth power (252^5 would exceed the capacity of a 12-digit calculator). Explain that the basic trick — extracting the fifth root through 99^5 — was too easy, and you'd now like more of a challenge (actually, the expanded version is no more difficult).

With this expanded version, you obtain the ones digit of the answer in the same manner as explained above. However, you then must solve for the hundreds and tens digits simultaneously.

The trick is to ignore the eight digits to the left of the ones digit, and concentrate on what remains. For example, let's say your subject takes his or her chosen number to the fifth power and hands you the product 150,536,645,632. You know that the ones digit of the fifth root is 2. Then ignore the eight digits to the left of the ones digit, and concentrate on what remains: 150. You then proceed as follows:

If the remaining number is in the following range:		Then the hundreds and tens digits are:
10–16	⟶	10
16–24	⟶	11
24–36	⟶	12
36–52	⟶	13
52–75	⟶	14
75–104	⟶	15
104–140	⟶	16
140–188	⟶	17
188–244	⟶	18
244–320	⟶	19
320–400	⟶	20
400–510	⟶	21
510–630	⟶	22
630–790	⟶	23
790–970	⟶	24
970–999	⟶	25

Returning to our example above, the remaining number was 150. As you can see from our chart, 150 fits in the 140–188 range. Therefore, the hundreds and tens digits are 17, and the fifth root is 172.

Let's try one more. What is the fifth root of 616,132,666,368? Well, we know that the ones digit of the answer is 8. Ignoring the next eight digits, we are left with 616. Our chart shows us that 616 fits in the 510–630 range (hundreds and tens digits = 22), so our answer is 228.

Now it's your turn to extract these "expanded version" fifth roots:

A.	10,000,000,000	E.	448,816,553,824	I.	98,465,804,768
B.	48,261,724,457	F.	312,079,600,999	J.	796,262,400,000
C.	59,797,108,943	G.	12,762,815,625	K.	21,924,480,357
D.	168,874,213,376	H.	539,218,609,632	L.	996,250,626,251

(See solutions on page 221.)

Week 4 Quick Quiz

Let's see how many tricks from Week 4 you can remember and apply by taking this brief test. There's no time limit, but try to work through these items as rapidly as possible. Before you begin, glance at the computations and try to identify the trick that you could use. In some instances, however, you will be asked to perform a calculation in a certain manner. When you flip ahead to the solutions, you will see which trick was intended.

Elementary Exercises

1. $25 \times 35 =$

2. $35 \times 15 =$

3. $\begin{array}{r} 204 \\ \times\ 505 \\ \hline \end{array}$

4. $10\frac{4}{5} \times 5 =$

5. Carry digits "below the line":

   $\begin{array}{r} 258 \\ 641 \\ 337 \\ +\ 109 \\ \hline \end{array}$

6. Break apart one number:

   $165 + 46 =$

7. $22 \times 82 =$

8. $74 \times 34 =$

9. $\frac{3}{4} \times 64 =$

10. $\begin{array}{r} 712 \\ \times\ \ \ 5 \\ \hline \end{array}$

11. $45 \times 111 =$

12. $231 \times 111 =$

13. $750 \times 999 =$

Brain Builders

1. $105 \times 95 =$

2. $\begin{array}{r} 703 \\ \times\ 406 \\ \hline \end{array}$

3. $27 \times 2\frac{8}{9} =$

4. $88 \times 19 =$

5. $\frac{5}{7} \times 280 =$

6. $62 \times 1{,}111 =$

7. $999 \times 33 =$

(See solutions on page 219.)

Number Potpourri 19

Quick Trick

When multiplying a number by 11, you can easily perform the calculation "all at once." Let's try 635 × 11:

$$\begin{array}{r} 635 \\ \times\ 11 \\ \hline 5 \end{array} \rightarrow \begin{array}{r} \scriptstyle 5 \\ 635 \\ \times\ 11 \\ \hline 85 \end{array} \rightarrow \begin{array}{r} \scriptstyle 3 \\ 635 \\ \times\ 11 \\ \hline 6{,}985 \end{array}$$

As you can see, we took 11 × 5 = 55 (write 5, carry 5). Then 11 × 3 = 33, plus 5 carried = 38 (write 8, carry 3). Finally, 11 × 6 = 66, plus 3 carried = 69.

So we multiplied in the usual way, except that we took care of the tens and ones digits (the entire 11) all at once. Try using this trick on 635 × 12. You'll see that it does work, but that a little more brain power is needed to pull off the execution.

Days 29 and 30
Checking Your Answer

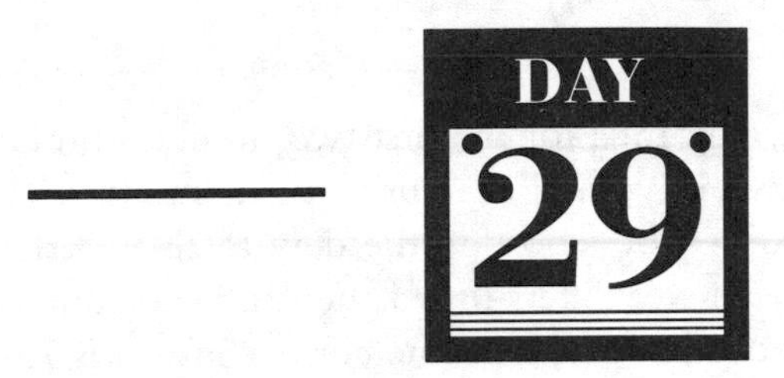

Trick 53: Checking Addition and Subtraction

Strategy: In the original *Rapid Math Tricks and Tips*, a checking device called "casting out nines" was explained. A similar technique, called "casting out elevens," will now be illustrated. Essentially, you work from right to left, subtracting digits, then adding them, as shown below. By themselves, casting out nines and elevens can definitely tell you if you have the wrong answer. However, when things work out correctly, each can only tell you that you probably (but not definitely) have the correct answer. However, when both methods indicate you probably have the correct answer, it is extremely probable that you do.

Elementary Example 1

427	←	$(7 - 2 + 4 = \mathbf{9})$	→	**9**
183	←	$(3 - 8 + 1 = -4; -4 + 11 = \mathbf{7})$	→	**7**
905	←	$(5 - 0 + 9 = 14; 14 - 11 = \mathbf{3})$	→	**3**
+ 664	←	$(4 - 6 + 6 = \mathbf{4})$	→	**+ 4**
2,179	←	$(9 - 7 + 1 - 2 = \mathbf{1})$		$23 - 11 - 11 = \mathbf{1}$
		↑ compare ↑		

Explanation: For each number, we begin with the ones digit, then subtract the tens digit, then add the hundreds digit. A "digit total" from 0 to 10 is allowable. However, when the digit total is negative, we add 11, and keep adding 11, if necessary, until we are within the 0–10 range. Similarly, when the digit total exceeds 10, we deduct 11, and keep deducting 11, if necessary, until we are within the 0–10 range. In our example above, we added the digit totals and arrived at a sum of 23. Because 23 exceeds 10, we subtract 11 twice to produce an allowable digit total of 1. We compare this digit total of 1 with the digit total of the answer. Because they agree (at 1), then we probably added correctly in the first place.

Elementary Example 2

27,185	←	$(5 - 8 + 1 - 7 + 2 = -7; -7 + 11 = \mathbf{4})$	→	**4**
− 19,340	←	$(0 - 4 + 3 - 9 + 1 = -9; -9 + 11 = \mathbf{2})$	→	**− 2**
7,745	←	$(5 - 4 + 7 - 7 = \mathbf{1})$		**2**

↑ compare ↑ (1 and 2)

Explanation: Calculate the digit total for the first two numbers (notice that they initially come out negative, so we add 11 to bring them into the allowable 0–10 range). Then we perform subtraction on the digit totals to arrive at 2. Because the digit total of the answer differs (it is 1, not 2), we definitely have the wrong answer to the subtraction problem (the correct answer is 7,845).

Brain Builder

25,386	←	$(6 - 8 + 3 - 5 + 2 = -2; -2 + 11 = \mathbf{9})$	→	**9**
90,447	←	$(7 - 4 + 4 - 0 + 9 = 16; 16 - 11 = \mathbf{5})$	→	**5**
+ 38,122	←	$(2 - 2 + 1 - 8 + 3 = -4; -4 + 11 = \mathbf{7})$	→	**+ 7**
153,955	←	$(5 - 5 + 9 - 3 + 5 - 1 = \mathbf{10})$		21
				− 11
				10

↑ compare ↑ (10 and 10)

Explanation: Calculate the digit total for each number (notice in one case we had to deduct 11 to be in the allowable range). Then add the digit totals, which equal 21 (9 + 5 + 7). Subtract 11 from the 21 and you're at 10. Because the digit total of the answer is also 10, we probably added correctly in the first place.

Food for Thought: When performing the alternate subtractions and additions, it is probably easier to first add all the alternate numbers, then add the other set of alternate numbers. Subtract one sum from the other, and you've got your digit total (unless, of course, you have to adjust up or down by elevens).

Practice Problems

Cast out elevens to determine if these answers are probably correct or definitely incorrect.

Elementary Exercises

1.
```
   531
   807   Definitely
   166   incorrect
 + 395
 1,889
```

2.
```
   337
   141
   786
 + 555
 1,819
```

3.
```
   806
   340
   651
 + 299
 2,096
```

4. $$\begin{array}{r} 916 \\ 452 \\ 660 \\ +\ 875 \\ \hline 2,093 \end{array}$$

5. $$\begin{array}{r} 411 \\ 278 \\ 956 \\ +\ 103 \\ \hline 1,648 \end{array}$$

6. $$\begin{array}{r} 678 \\ 192 \\ 547 \\ +\ 365 \\ \hline 1,782 \end{array}$$

7. $$\begin{array}{r} 101 \\ 736 \\ 444 \\ +\ 892 \\ \hline 2,273 \end{array}$$

8. $$\begin{array}{r} 43,792 \\ -\ 16,804 \\ \hline 26,988 \end{array}$$

9. $$\begin{array}{r} 76,205 \\ -\ 29,558 \\ \hline 47,647 \end{array}$$

10. $$\begin{array}{r} 51,893 \\ -\ 36,051 \\ \hline 15,842 \end{array}$$

11. $$\begin{array}{r} 19,211 \\ -\ 13,865 \\ \hline 5,436 \end{array}$$

12. $$\begin{array}{r} 24,719 \\ -\ 8,446 \\ \hline 16,273 \end{array}$$

13. $$\begin{array}{r} 98,170 \\ -\ 62,342 \\ \hline 35,828 \end{array}$$

14. $$\begin{array}{r} 70,496 \\ -\ 35,877 \\ \hline 34,419 \end{array}$$

Brain Builders

1. $$\begin{array}{r} 42,775 \\ 37,104 \\ +\ 70,336 \\ \hline 150,215 \end{array}$$

2. $$\begin{array}{r} 81,363 \\ 44,009 \\ +\ 19,775 \\ \hline 145,137 \end{array}$$

3. $$\begin{array}{r} 62,107 \\ 34,658 \\ +\ 81,542 \\ \hline 148,307 \end{array}$$

4.
$$\begin{array}{r} 10{,}574 \\ 76{,}053 \\ +\ 54{,}610 \\ \hline 141{,}237 \end{array}$$

5.
$$\begin{array}{r} 365{,}401 \\ -\ 188{,}656 \\ \hline 176{,}755 \end{array}$$

6.
$$\begin{array}{r} 807{,}361 \\ -\ 554{,}088 \\ \hline 253{,}273 \end{array}$$

7.
$$\begin{array}{r} 660{,}512 \\ -\ 325{,}469 \\ \hline 335{,}043 \end{array}$$

8.
$$\begin{array}{r} 204{,}978 \\ -\ 96{,}283 \\ \hline 108{,}965 \end{array}$$

(See solutions on page 216.)

Number Potpourri 20

Multiplication Madness

Pick a multiplication problem, any multiplication problem. How about 21 × 34? Then successively cut the left-hand number in half while doubling the right-hand number until the left-hand number is reduced to 1. Whenever a result is a mixed number, round down. For example, 5 divided by 2 is written not as $2\frac{1}{2}$, but as 2.

When you're finished, cross out each row in which the left-hand number is even. Then add the remaining right-hand numbers, and you've got the product. Let's try this bizarre technique on our 21 × 34 example:

21 × 34
~~10 × 68~~
5 × 136
~~2 × 272~~
1 × 544
714 (= 21 × 34)

It's hard to believe, but this trick will work on any multiplication problem. Try it yourself!

DAY 30

Trick 54: Checking Multiplication and Division

Strategy: We are going to check multiplication and division using the same technique we used in Trick 53. So make sure you fully understand that trick before proceeding to this one. Again, we are going to obtain digit totals by casting out elevens. However, we are going to multiply digit totals, rather than add or subtract them, and then compare the result with the digit total of the answer we are checking. If the two digit totals agree, you probably have the correct answer. However, if they do not agree, then you definitely have the wrong answer. Remember to add or deduct 11 (or multiples of 11) if any digit total is not within the acceptable range of 0–10. As you'll see in Elementary Example 2 below, we are going to convert our division into a multiplication and proceed as in Elementary Example 1.

Elementary Example 1

459 × 156

459	←	(9 − 5 + 4 = **8**) ⟶	**8**
× 156	←	(6 − 5 + 1 = **2**) ⟶	× **2**
71,604	←	(4 − 0 + 6 − 1 + 7 = 16; 16 − 11 = **5**)	16 − 11 = **5**
		↑ compare	↑

Explanation: Basically, we obtained digit totals for the two factors (8 and 2) and multiplied them together. The resulting digit total (after having to deduct 11) is 5. Because that digit total equals that of the answer we are checking, we probably multiplied correctly in the first place.

Elementary Example 2

209,346 ÷ 222

953	←	(3 − 5 + 9 = **7**) ⟶	**7**
× 222	←	(2 − 2 + 2 = **2**) ⟶	× **2**
209,346	←	(6 − 4 + 3 − 9 + 0 − 2 = −6; −6 + 11 = **5**)	14
			− 11
			3
		↑ compare	↑

Explanation: Notice that we converted the division problem into a multiplication and checked as in Elementary Example 1. Because the resulting digit totals of 5 and 3 do not agree, we definitely did not divide correctly in the first place.

Brain Builder

6,724 × 3,198

$$\begin{array}{rl}
6{,}724 \leftarrow & (4 - 2 + 7 - 6 = \mathbf{3}) \longrightarrow \quad 3 \\
\times\, 3{,}198 \leftarrow & (8 - 9 + 1 - 3 = -3;\ -3 + 11 = \mathbf{8}) \longrightarrow \quad \times\, 8 \\
\hline
21{,}503{,}352 \leftarrow & (2 - 5 + 3 - 3 + 0 - 5 + 1 - 2 = -9;\ -9 + 11 = \mathbf{2}) \quad 24 \\
 & \qquad -11 \\
 & \qquad -11 \\
 & \qquad \mathbf{2} \\
 & \text{compare}
\end{array}$$

Explanation: Once again, we obtained digit totals, multiplied, and compared results. Because the two resulting digit totals of 2 agree, then we probably multiplied correctly in the first place.

Food for Thought: When dividing, you will sometimes obtain a remainder. Just apply the technique above (converting to a multiplication), ignore the remainder, and you should be off by the exact amount of the remainder. To simplify our explanation, however, we purposely chose a division problem without a remainder.

Practice Problems

Cast out elevens to determine if these answers are probably correct or definitely incorrect.

Elementary Exercises

1. $\begin{array}{r} 216 \\ \times\, 743 \\ \hline 160{,}488 \end{array}$ **Probably correct**

2. $\begin{array}{r} 406 \\ \times\, 831 \\ \hline 337{,}386 \end{array}$

3. $\begin{array}{r} 336 \\ \times\, 590 \\ \hline 198{,}420 \end{array}$

4. $\begin{array}{r} 118 \\ \times\, 734 \\ \hline 86{,}612 \end{array}$

5. $\begin{array}{r} 681 \\ \times\ 402 \\ \hline 243{,}762 \end{array}$

6. $\begin{array}{r} 737 \\ \times\ 128 \\ \hline 94{,}346 \end{array}$

7. $\begin{array}{r} 417 \\ \times\ 922 \\ \hline 384{,}474 \end{array}$

8. $145{,}200 \div 825 = 175$

9. $58{,}366 \div 154 = 379$

10. $149{,}655 \div 907 = 165$

11. $293{,}468 \div 329 = 982$

12. $295{,}704 \div 888 = 313$

13. $132{,}506 \div 634 = 209$

14. $400{,}026 \div 551 = 726$

Brain Builders

1. $\begin{array}{r} 8{,}254 \\ \times\ 3{,}195 \\ \hline 26{,}171{,}530 \end{array}$

2. $\begin{array}{r} 1{,}095 \\ \times\ 8{,}774 \\ \hline 9{,}607{,}530 \end{array}$

3. $\begin{array}{r} 6{,}227 \\ \times\ 4{,}335 \\ \hline 26{,}994{,}045 \end{array}$

4. $\begin{array}{r} 9{,}820 \\ \times\ 5{,}643 \\ \hline 55{,}414{,}270 \end{array}$

5. $1{,}682{,}184 \div 3{,}192 = 527$

6. $5{,}396{,}186 \div 5{,}407 = 999$

7. $2{,}793{,}686 \div 6{,}881 = 406$

8. $1{,}318{,}260 \div 3{,}460 = 341$

(See solutions on page 217.)

Number Potpourri 21

For the Bold and the Brilliant

Here are a couple of math tricks that are extremely interesting. Unfortunately, very few people can execute them efficiently in their heads.

The first trick involves multiplication by 101, 202, 303, and so forth. The secret is to multiply by the nearest multiple of 100, then add 1%. For example, $67 \times 404 = 67 \times 400$, plus 1% of that product. This turns out to equal 26,800 + 268, or 27,068.

The second trick is very similar to the first one and involves multiplication by 99 and multiples of 99 (i.e., 198, 297, etc.). The secret is to multiply by the nearest multiple of 100, then subtract 1%. For example, $84 \times 297 = 84 \times 300$, minus 1% of that product. This turns out to equal 25,200 – 252, or 24,948.

Want a final challenge? See if you can figure out a trick (similar to those shown above) to multiply by 111, 222, 333, etc. The flip side of that trick involves multiplication by 89, and multiples of 89 (i.e., 178, 267, etc.). If you can figure out these two tricks (and actually work them in your head), you're well on your way to becoming a certified mental math monster!

Final Exam

Well, this is it! Take a deep breath and relax. When you finish, grade your own test (be honest now) and then go and celebrate. Use any rapid math technique you like, unless otherwise indicated. You will see which trick was intended when you flip ahead to the solutions. Ready, set, go!

1. Multiply without carrying:

$$\begin{array}{r} 39 \\ \times\ 7 \\ \hline \end{array}$$

2. Add without carrying:

$$\begin{array}{r} 491 \\ 725 \\ 306 \\ +\ 668 \\ \hline \end{array}$$

3.

$$\begin{array}{r} 470 \\ -\ 235 \\ \hline \end{array}$$

4. $132 \div 12 =$

5.

$$\begin{array}{r} 39 \\ -\ 17 \\ +\ 20 \\ +\ 44 \\ -\ 47 \\ \hline \end{array}$$

6.

$$\begin{array}{r} 19 \\ 22 \\ 21 \\ 16 \\ +\ 18 \\ \hline \end{array}$$

7. Multiply by factors:

$$\begin{array}{r} 294 \\ \times\ 63 \\ \hline \end{array}$$

8. $4 \times 7 \times 25 \times 3 =$

9. $24 \times 6 =$

10. $114 \div 6 =$

11.

$$\begin{array}{r} 43 \\ 47 \\ 15 \\ 49 \\ 44 \\ +\ 42 \\ \hline \end{array}$$

12.

$$\begin{array}{r} \$6.95 \\ 3.98 \\ 4.92 \\ 5.90 \\ +\ 1.97 \\ \hline \end{array}$$

13.

$$\begin{array}{r} 35 \\ \times\ 42 \\ \hline \end{array}$$

14. $8\overline{)2,856}$

15. Divide by two one-digit divisors:

    $3{,}465 \div 45 =$

16. Multiply the tens digit first:

$$\begin{array}{r} 63 \\ \times\ 7 \\ \hline \end{array}$$

17. Multiply the hundreds digit first:

$$\begin{array}{r} 218 \\ \times\ \ 6 \\ \hline \end{array}$$

18. $116 \div 4 =$

19. $378 \div 9 =$

20. $16 \times 31 =$

21. $7 \times 49 =$

22. $14 \times 12 =$

23. $300 \div 12 =$

24. $495 \div 5 =$

25. $413 \div 7 =$

26. $358 - 193 =$

27. $18 \times 14 =$

28. $31^2 =$

29. $18 \times 18 =$

30. $23^2 =$

31. $9 \times 44 =$

32. $8 \times 45 =$

33. $36^2 =$

34. $24 \times 24 =$

35. Multiply from left to right:

$$\begin{array}{r} 5{,}314 \\ \times\ \ \ \ 6 \\ \hline \end{array}$$

36. Do all the work in your head:

$$\begin{array}{r} 35 \\ \times\ 14 \\ \hline \end{array}$$

37. $98^2 =$

38. $109 \times 109 =$

39. $$\begin{array}{r} 97 \\ \times\ 95 \\ \hline \end{array}$$

40. $45 \times 35 =$

41. $25 \times 45 =$

42. 705×204

43. $80 \times 3\frac{3}{4} =$

44. Carry digits "below the line":

$$\begin{array}{r} 546 \\ 109 \\ 832 \\ +\ 276 \\ \hline \end{array}$$

45. Break apart one number:

$188 + 36 =$

46. $33 \times 64 =$

47. $76 \times 36 =$

48. $\frac{4}{5} \times 75 =$

49. 516×4

50. $72 \times 111 =$

51. $423 \times 111 =$

52. $640 \times 999 =$

For Questions 53 and 54, cast out elevens and state whether each answer is "probably correct" or "definitely incorrect."

53.

$$\begin{array}{r} 365 \\ 881 \\ 407 \\ +\ 692 \\ \hline 2{,}355 \end{array}$$

54.

$$\begin{array}{r} 518 \\ \times\ 724 \\ \hline 375{,}032 \end{array}$$

(See solutions on page 220.)

Number Challenge 5

Using just three digits and no additional mathematical symbols, how would you represent the largest number possible?

(See answer on page 222.)

— Conclusion —

Congratulations — you made it! You are well on your way to becoming a mental math wizard. In the meantime, let me make a couple of final recommendations:

- Take some time to determine which of the 54 tricks you find the most useful, most interesting, and easiest to remember. Then focus on those tricks in your daily life, and don't worry so much about the others. Now and then, reexamine all 54 to modify your working repertoire. But most important, practice these tricks every chance you get!
- Keep your mind "mathematically fit." When you go to the supermarket, estimate the total cost of groceries before they are rung up at the checkout counter. Your estimate should be within 5% of the actual total. Whenever you see numbers on a license plate, billboard, or anyplace else, manipulate those numbers in some way.

Haven't had enough? Want to learn *all* the tricks of the trade? Well, don't forget about the original *Rapid Math Tricks and Tips* book, which can be purchased (or ordered) from any bookstore. I can personally vouch for the book's author.

In any case, I hope you enjoyed your excursion into the world of rapid calculation. For those interested, I hold rapid math workshops on a fairly regular basis. If you would like me to address your organization, school, or business, please write to me at California Lutheran University, c/o School of Business, 60 West Olsen Road, Thousand Oaks, California 91360, for more information.

I would also enjoy receiving your comments, suggestions, or perhaps your very own rapid math techniques. Please write to me at the above address. I look forward to hearing from you.

Thank you, and good luck!

Edward H. Julius

Appendix A: Tricks from *Rapid Math Tricks and Tips* That You Need to Know for This Book

(Abridgments)

1: Rapidly Square Any Number Ending in 5

Strategy: This trick is one of the oldest in the book, and one of the best! To square a number that ends in 5, first multiply the tens digit by the next whole number. To that product, affix the number 25. The number to affix (25) is easy to remember, because $5^2 = 25$. Read on to see how this marvelous trick works.

Elementary Example 1

15^2

Step 1. Multiply: $1 \times 2 = 2$.

Step 2. Affix 25: 225 (answer).

Thought Process Summary

$$\begin{array}{r} 15 \\ \times\,15 \\ \hline \end{array} \rightarrow \begin{array}{r} 1 \\ \times\,2 \\ \hline 2 \end{array} \rightarrow 225$$

Elementary Example 2

65^2

Step 1. Multiply: $6 \times 7 = 42$.

Step 2. Affix 25: 4,225 (answer).

Thought Process Summary

$$\begin{array}{r} 65 \\ \times\,65 \\ \hline \end{array} \rightarrow \begin{array}{r} 6 \\ \times\,7 \\ \hline 42 \end{array} \rightarrow 4{,}225$$

2: Rapidly Multiply Any Two-Digit Number by 11

Strategy: The 11 trick is the most popular trick of them all — and one of the must useful. To multiply a two-digit number by 11, first write the number, leaving some space between the two digits. Then insert the sum of the number's two digits in-between the two digits. You will have to carry when the sum of the digits exceeds 9. This trick is almost too good to be true, as you'll see in the following examples.

Elementary Example 1

24 × 11

Step 1. Write the multiplicand in the answer space, leaving some room between the two digits: 2 ? 4.

Step 2. Add the two digits of the multiplicand: 2 + 4 = 6.

Step 3. Insert the 6 between the two digits (see Step 1), producing the answer 264.

Thought Process Summary

24 × 11 → 2 ? 4 → 2 + 4 = 6 → 264

Elementary Example 2

76 × 11

Step 1. Write the multiplicand in the answer space, leaving some room between the two digits: 7 ? 6.

Step 2. Add the two digits of the multiplicand: 7 + 6 = 13.

Step 3. Insert the ones digit of the 13 between the two digits (see Step 1) producing an intermediary product of 736.

Step 4. Because the two digits of the multiplicand total more than 9, one must carry, converting 736 to the answer, 836.

Thought Process Summary

$$\begin{array}{r} 76 \\ \times\ 11 \\ \hline \end{array} \rightarrow 7\ ?\ 6 \rightarrow \begin{array}{r} 7 \\ +\ 6 \\ \hline 13 \end{array} \rightarrow 736 \rightarrow 836$$

Appendix B: List of the 60 Tricks Covered in *Rapid Math Tricks and Tips*

1. Multiplying and Dividing with Zeros
2. Multiplying and Dividing with Decimal Points
3. Rapidly Multiply by 4
4. Rapidly Divide by 4
5. Rapidly Multiply by 5
6. Rapidly Divide by 5
7. Rapidly Square Any Number Ending in 5
8. Rapidly Multiply Any Two-Digit Number by 11
9. Rapidly Multiply by 25
10. Rapidly Divide by 25
11. Rapidly Multiply Any One- or Two-Digit Number by 99
12. Rapidly Multiply Any One- or Two-Digit Number by 101
13. Rapidly Multiply Two Numbers Whose Difference Is 2
14. Rapidly Check Multiplication and Division
15. Rapidly Multiply by 125
16. Rapidly Divide by 125
17. Rapidly Multiply by 9
18. Rapidly Multiply by 12
19. Rapidly Multiply by 15
20. Rapidly Multiply Two Numbers with a Special Relationship
21. Rapidly Multiply by 1.5, 2.5, 3.5, etc.
22. Rapidly Divide by 1.5, 2.5, 3.5, etc.
23. Rapidly Square Any Two-Digit Number Beginning in 5
24. Rapidly Square Any Two-Digit Number Ending in 1
25. Rapidly Multiply Two-Digit Numbers without Showing Work
26. Rapidly Multiply Two Numbers Whose Difference Is 4
27. Rapidly Multiply in Two Steps
28. Rapidly Multiply Two Numbers That Are Just Over 100
29. Rapidly Subtract by Adding
30. Rapidly Subtract by Adding — A Variation
31. Rapidly Subtract by Altering
32. Rapidly Add by Altering
33. Rapidly Add by Grouping and Reordering
34. Rapidly Add without Carrying

35. Rapidly Add a Column of Numbers — Variation 1
36. Rapidly Add a Column of Numbers — Variation 2
37. Rapidly Add with the "Cross-Out" Technique
38. Rapidly Add Columns of Numbers in Sections
39. Rapidly Add a Few Numbers
40. Rapidly Add 1 + 2 + 3, etc.
41. Rapidly Subtract in Two Steps
42. Rapidly Check Addition and Subtraction
43. Rapidly Multiply by 75
44. Rapidly Divide by 75
45. Rapidly Divide by 8
46. Rapidly Divide by 15
47. Rapidly Estimate Multiplication by 33 or 34
48. Rapidly Estimate Division by 33 or 34
49. Rapidly Estimate Multiplication by 49 or 51
50. Rapidly Estimate Division by 49 or 51
51. Rapidly Estimate Multiplication by 66 or 67
52. Rapidly Estimate Division by 66 or 67
53. Rapidly Estimate Division by 9
54. Rapidly Estimate Division by 11
55. Rapidly Estimate Division by 14
56. Rapidly Estimate Division by 17
57. Rapidly Multiply by Regrouping
58. Rapidly Multiply by Augmenting
59. Rapidly Multiply Any Three-Digit or Larger Number by 11
60. Rapidly Divide by 9, 99, 999, etc.

Bonus: Parlor Tricks

1. Finding the Fifth Root
2. The Perpetual Calendar
3. Multiplying Two 3-Digit Numbers without Showing Work
4. Rapidly Subtract the Squares of Any Two Numbers
5. "I'd Like to Speak to the Wizard"

— Solutions —

Exercises

Trick 1

Elementary

1. 368
2. 465
3. 236
4. 693
5. 588
6. 588
7. 390
8. 747
9. 224
10. 196
11. 344
12. 474
13. 477
14. 195

Brain Builders

1. 2,982
2. 4,590
3. 2,992
4. 8,973
5. 3,870
6. 3,556
7. 1,883
8. 5,928

Trick 2

Elementary

1. 1,895
2. 2,265
3. 2,292
4. 1,657
5. 1,989
6. 2,296
7. 2,412
8. 1,829
9. 2,253
10. 2,156
11. 2,767
12. 2,664
13. 1,981
14. 2,679

Brain Builders

1. 35,391
2. 28,952
3. 27,477
4. 26,281
5. 35,976
6. 29,533
7. 33,251
8. 33,706

Trick 3

Elementary

1. 114
2. 327
3. 152
4. 277
5. 505
6. 163
7. 225
8. 326
9. 292
10. 54
11. 615
12. 363
13. 191
14. 407

Brain Builders

1. 3,016
2. 1,242
3. 2,448
4. 7,156
5. 5,514
6. 16,250
7. 32,235
8. 24,125

Trick 4

Elementary

1. 5
2. 8
3. 8
4. 4
5. 6
6. 7
7. 5
8. 5
9. 11
10. 7
11. 8
12. 11
13. 3
14. 12

Brain Builders

1. 9
2. 11
3. 15
4. 13
5. 14
6. 12
7. 7
8. 8

Trick 5

Elementary

1. 24
2. 42
3. 78
4. 55
5. 38
6. 47
7. 26
8. 44
9. 12
10. 56
11. 50
12. 68
13. 22
14. 44

Brain Builders

1. 69
2. 0
3. 70
4. 97
5. 74
6. 59
7. 66
8. 33

Trick 6

Elementary

1. 162
2. 121
3. 317
4. 198
5. 81
6. 479
7. 236
8. 643
9. 398
10. 718
11. 560
12. 161
13. 282
14. 484

Brain Builders

1. 703
2. 4,199
3. 1,048
4. 6,296
5. 2,800
6. 4,906
7. 1,752
8. 7,001

Trick 7

Elementary

1. 2,160
2. 16,023
3. 7,532
4. 41,104
5. 37,296
6. 11,886
7. 15,008
8. 17,199
9. 12,480
10. 20,142
11. 17,928
12. 27,936
13. 30,780
14. 12,475

Brain Builders

1. 87,010
2. 92,394
3. 335,160
4. 309,771
5. 40,488
6. 246,384
7. 237,456
8. 559,944

Trick 8

Elementary

1. 350
2. 540
3. 330
4. 490
5. 1,800
6. 480
7. 630
8. 240
9. 2,100
10. 360
11. 600
12. 180
13. 900
14. 1,300

Brain Builders

1. 77
2. 54
3. 170
4. 49
5. 120
6. 160
7. 55
8. 63

Trick 9

Elementary

1. 108
2. 132
3. 78
4. 192
5. 162
6. 96
7. 138
8. 90
9. 114
10. 210
11. 198
12. 102
13. 144
14. 84

Brain Builders

1. 330
2. 204
3. 156
4. 510
5. 216
6. 252
7. 390
8. 222

Trick 10

Elementary

1. 35
2. 15
3. 18
4. 32
5. 17
6. 23
7. 19
8. 16
9. 24
10. 13
11. 33
12. 14
13. 27
14. 22

Brain Builders

1. 55
2. 130
3. 75
4. 36
5. 85
6. 34
7. 140
8. 65

Trick 11

Elementary

1. 406
2. 679
3. 462
4. 228
5. 721
6. 418
7. 338
8. 435
9. 414
10. 365
11. 355
12. 446
13. 487
14. 550

Brain Builders

1. 3,573
2. 4,281
3. 2,575
4. 4,028
5. 6,274
6. 3,323
7. 3,227
8. 4,787

Trick 12

Elementary

1. $19.78
2. $20.76
3. $21.75
4. $22.73
5. $23.76
6. $26.66
7. $23.88
8. $24.58
9. $20.75
10. $23.70
11. $22.72
12. $20.88
13. $24.74
14. $22.75

Brain Builders

1. $40.45
2. $37.50
3. $41.36
4. $34.50
5. $35.58
6. $34.48
7. $41.50
8. $37.51

Trick 13

Elementary

1. 2,730
2. 775
3. 756
4. 576
5. 2,079
6. 1,512
7. 1,953
8. 2,736
9. 1,845
10. 992
11. 2,688
12. 1,536
13. 1,476
14. 1,560

Brain Builders

1. 9,262
2. 3,675
3. 26,944
4. 5,040
5. 4,340
6. 2,835
7. 6,804
8. 44,880

Trick 14

Elementary

1. 347
2. 489
3. 166
4. 252
5. 834
6. 619
7. 972
8. 508
9. 764
10. 388
11. 137
12. 999
13. 787
14. 258

Brain Builders

1. 244
2. 541
3. 972
4. 345
5. 416
6. 177
7. 356
8. 215

Trick 15

Elementary

1. 74
2. 83
3. 27
4. 65
5. 89
6. 71
7. 42
8. 33
9. 19
10. 68
11. 53
12. 97
13. 29
14. 34

Brain Builders

1. 749
2. 361
3. 805
4. 198
5. 977
6. 435
7. 526
8. 284

Trick 16

Elementary

1. 192
2. 215
3. 105
4. 111
5. 162
6. 196
7. 162
8. 238
9. 360
10. 228
11. 144
12. 170
13. 220
14. 171

Brain Builders

1. 435
2. 504
3. 399
4. 380
5. 648
6. 354
7. 335
8. 752

Trick 17

Elementary

1. 492
2. 2,145
3. 678
4. 1,890
5. 1,001
6. 1,028
7. 992
8. 2,355
9. 1,665
10. 1,386
11. 1,662
12. 956
13. 1,659
14. 2,475

Brain Builders

1. 5,684
2. 3,460
3. 3,498
4. 3,688
5. 6,579
6. 2,082
7. 4,456
8. 3,740

Trick 18

Elementary

1. 23
2. 56
3. 29
4. 25
5. 57
6. 33
7. 42
8. 32
9. 58
10. 53
11. 63
12. 107
13. 152
14. 87

Brain Builders

1. 23
2. 45
3. 22
4. 26
5. 21
6. 24
7. 24
8. 66

Trick 19

Elementary

1. 35
2. 23
3. 24
4. 16
5. 33
6. 52
7. 22
8. 35
9. 46
10. 42
11. 57
12. 43
13. 71
14. 64

Brain Builders

1. 21
2. 32
3. 14
4. 63
5. 42
6. 45
7. 51
8. 32

Trick 20

Elementary

1. 945
2. 714
3. 744
4. 852
5. 1,025
6. 976
7. 1,365
8. 2,430
9. 403
10. 1,122
11. 924
12. 1,952
13. 697
14. 2,835

Brain Builders

1. 1,785
2. 4,473
3. 2,255
4. 3,172
5. 2,883
6. 5,265
7. 4,284
8. 6,552

Trick 21

Elementary

1. 133
2. 392
3. 174
4. 395
5. 531
6. 396
7. 273
8. 267
9. 345
10. 392
11. 171
12. 354
13. 203
14. 156

Brain Builders

1. 588
2. 2,670
3. 759
4. 1,580
5. 585
6. 456
7. 2,360
8. 377

Trick 22

Elementary

1. 180
2. 108
3. 156
4. 204
5. 264
6. 192
7. 276
8. 228
9. 312
10. 216
11. 288
12. 168
13. 252
14. 300

Brain Builders

1. 420
2. 384
3. 900
4. 624
5. 540
6. 324
7. 660
8. 432

Trick 23

Elementary

1. 13
2. 17
3. 14
4. 9
5. 23
6. 16
7. 24
8. 25
9. 15
10. 21
11. 18
12. 8
13. 19
14. 22

Brain Builders

1. 45
2. 27
3. 75
4. 36
5. 32
6. 35
7. 26
8. 65

Trick 24

Elementary

1. 18
2. 49
3. 23
4. 97
5. 48
6. 37
7. 149
8. 72
9. 58
10. 196
11. 98
12. 77
13. 247
14. 99

Brain Builders

1. 19
2. 18
3. 38
4. 19
5. 27
6. 37
7. 18
8. 59

Trick 25

Elementary

1. 48
2. 19
3. 29
4. 18
5. 39
6. 29
7. 48
8. 97
9. 58
10. 49
11. 57
12. 99
13. 69
14. 78

Brain Builders

1. 9
2. 48
3. 39
4. 99
5. 48
6. 19
7. 37
8. 97

Trick 26

Elementary

1. 53
2. 169
3. 158
4. 246
5. 184
6. 224
7. 138
8. 367
9. 44
10. 153
11. 176
12. 126
13. 437
14. 171

Brain Builders

1. 550
2. 347
3. 378
4. 542
5. 524
6. 373
7. 518
8. 646

Trick 27

Elementary

1. 180
2. 198
3. 256
4. 221
5. 168
6. 285
7. 209
8. 252
9. 228
10. 238
11. 196
12. 208
13. 234
14. 270

Brain Builders

1. 361
2. 272
3. 323
4. 324
5. 304
6. 342
7. 306
8. 289

Trick 28

Elementary

1. 961
2. 841
3. 2,601
4. 1,521
5. 2,401
6. 3,721
7. 361
8. 1,681
9. 441
10. 3,481
11. 3,721
12. 2,401
13. 841
14. 961

Brain Builders

1. 6,241
2. 8,281
3. 4,761
4. 6,561
5. 5,041
6. 7,921
7. 9,801
8. 10,201

Trick 29

Elementary

1. 784
2. 1,024
3. 1,444
4. 2,704
5. 3,844
6. 2,304
7. 324
8. 1,764
9. 484
10. 3,364
11. 3,844
12. 2,304
13. 1,024
14. 784

Brain Builders

1. 6,724
2. 4,624
3. 8,464
4. 6,084
5. 5,184
6. 7,744
7. 9,604
8. 10,404

Trick 30

Elementary

1. 529
2. 729
3. 1,369
4. 2,809
5. 3,969
6. 2,209
7. 289
8. 1,849
9. 1,369
10. 1,089
11. 169
12. 3,249
13. 2,209
14. 3,969

Brain Builders

1. 6,889
2. 4,489
3. 8,649
4. 5,929
5. 5,329
6. 7,569
7. 9,409
8. 10,609

Trick 31

Elementary

1. 198
2. 462
3. 308
4. 792
5. 264
6. 330
7. 264
8. 352
9. 385
10. 176
11. 396
12. 385
13. 594
14. 462

Brain Builders

1. 770
2. 1,430
3. 2,310
4. 561
5. 1,320
6. 1,540
7. 3,960
8. 792

Trick 32

Elementary

1. 252
2. 315
3. 162
4. 648
5. 162
6. 153
7. 504
8. 405
9. 324
10. 126
11. 324
12. 216
13. 171
14. 360

Brain Builders

1. 702
2. 1,260
3. 1,080
4. 603
5. 2,430
6. 1,350
7. 540
8. 675

Trick 33

Elementary

1. 1,296
2. 3,136
3. 256
4. 2,116
5. 676
6. 256
7. 2,116
8. 1,296
9. 3,136
10. 676

Brain Builders

1. 7,396
2. 4,356
3. 9,216
4. 5,776
5. 4,356
6. 7,396
7. 5,776
8. 11,236

Trick 34

Elementary

1. 1,156
2. 2,916
3. 196
4. 1,936
5. 576
6. 196
7. 1,936
8. 1,156
9. 2,916
10. 576

Brain Builders

1. 7,056
2. 4,096
3. 8,836
4. 5,476
5. 4,096
6. 5,476
7. 7,056
8. 10,816

Trick 35

Elementary

1. 20,660
2. 26,000
3. 10,668
4. 19,845
5. 21,240
6. 16,212
7. 25,665
8. 9,872
9. 12,963
10. 22,624
11. 25,224
12. 12,720
13. 31,608
14. 12,112

Brain Builders

1. 22,035
2. 23,436
3. 68,754
4. 30,265
5. 33,004
6. 38,208
7. 6,534
8. 52,255

Trick 36

Elementary

1. 322
2. 2,236
3. 1,674
4. 924
5. 676
6. 1,248
7. 726
8. 510
9. 1,224
10. 476
11. 1,806
12. 875
13. 990
14. 1,815

Brain Builders

1. 3,358
2. 2,632
3. 2,867
4. 2,024
5. 2,028
6. 3,654
7. 1,188
8. 2,183

Trick 37

Elementary

1. 9,216
2. 9,801
3. 9,025
4. 8,281
5. 9,409
6. 8,649
7. 9,604
8. 8,464
9. 8,836
10. 9,801
11. 8,281
12. 9,216
13. 8,464
14. 9,025

Brain Builders

1. 994,009
2. 988,036
3. 982,081
4. 998,001
5. 986,049
6. 990,025
7. 984,064
8. 992,016

Trick 38

Elementary

1. 10,816
2. 10,201
3. 11,025
4. 11,881
5. 10,609
6. 11,449
7. 10,404
8. 11,664
9. 11,236
10. 10,201
11. 11,881
12. 10,816
13. 11,664
14. 11,025

Brain Builders

1. 1,006,009
2. 1,012,036
3. 1,018,081
4. 1,002,001
5. 1,014,049
6. 1,010,025
7. 1,016,064
8. 1,008,016

Trick 39

Elementary

1. 9,207
2. 8,832
3. 9,118
4. 8,360
5. 8,930
6. 8,633
7. 8,372
8. 9,306
9. 8,439
10. 8,640
11. 9,114
12. 8,170
13. 9,801
14. 7,990

Brain Builders

1. 991,018
2. 982,077
3. 990,016
4. 986,045
5. 983,052
6. 993,012
7. 980,075
8. 982,072

Trick 40

Elementary

1. 375
2. 2,475
3. 875
4. 1,575
5. 75
6. 375
7. 3,575
8. 2,475
9. 1,575
10. 3,575
11. 75
12. 875

Brain Builders

1. 4,875
2. 8,075
3. 6,375
4. 9,975
5. 8,075
6. 6,375
7. 9,975
8. 4,875

Trick 41

Elementary

1. 125
2. 1,925
3. 1,125
4. 2,925
5. 525
6. 1,125
7. 125
8. 525
9. 1,925
10. 2,925

Brain Builders

1. 4,125
2. 5,525
3. 7,125
4. 8,925
5. 7,125
6. 4,125
7. 8,925
8. 5,525

Trick 42

Elementary

1. 62,220
2. 53,424
3. 242,606
4. 42,435
5. 303,812
6. 403,106
7. 183,305
8. 122,412
9. 102,921
10. 124,236
11. 98,209
12. 243,808
13. 83,018
14. 53,607

Brain Builders

1. 547,828
2. 487,224
3. 359,156
4. 368,242
5. 249,981
6. 206,448
7. 428,542
8. 729,927

Trick 43

Elementary

1. 48
2. 69
3. 120
4. 600
5. 780
6. 87
7. 63
8. 540
9. 53
10. 560
11. 59
12. 570
13. 84
14. 700

Brain Builders

1. 58
2. 69
3. 52
4. 510
5. 78
6. 330
7. 84
8. 117

Trick 44

Elementary

1. 2,365
2. 2,397
3. 2,254
4. 2,711
5. 2,317
6. 2,863
7. 2,233
8. 2,370
9. 2,625
10. 2,925
11. 2,661
12. 3,306
13. 3,007
14. 2,883

Brain Builders

1. 30,776
2. 24,811
3. 28,675
4. 29,288
5. 32,717
6. 23,515
7. 29,292
8. 25,426

Trick 45

Elementary

1. 213
2. 133
3. 107
4. 241
5. 171
6. 223
7. 155
8. 130
9. 272
10. 141
11. 231
12. 214
13. 126
14. 234

Brain Builders

1. 522
2. 351
3. 434
4. 732
5. 701
6. 414
7. 920
8. 804

Trick 46

Elementary

1. 924
2. 2,530
3. 407
4. 4,004
5. 1,210
6. 5,412
7. 2,112
8. 1,045
9. 1,606
10. 308
11. 3,630
12. 2,024
13. 1,221
14. 3,025

Brain Builders

1. 8,008
2. 2,156
3. 4,554
4. 4,818
5. 1,463
6. 5,445
7. 7,216
8. 2,849

Trick 47

Elementary

1. 2,304
2. 2,925
3. 1,001
4. 2,016
5. 2,809
6. 2,736
7. 1,804
8. 2,501
9. 2,625
10. 1,209
11. 2,916
12. 3,036
13. 2,201
14. 2,125

Brain Builders

1. 3,381
2. 2,464
3. 3,481
4. 2,849
5. 1,649
6. 3,364
7. 2,581
8. 3,149

Trick 48

Elementary

1. 45
2. 27
3. 28
4. 48
5. 24
6. 38
7. 33
8. 60
9. 120
10. 105
11. 100
12. 120
13. 66
14. 150

Brain Builders

1. 35
2. 42
3. 56
4. 350
5. 200
6. $16\frac{1}{2}$
7. 420
8. 120

Trick 49

Elementary

1. 3,575
2. 4,599
3. 540
4. 488
5. 2,763
6. 5,680
7. 798
8. 1,569
9. 3,672
10. 872
11. 4,065
12. 345
13. 6,496
14. 4,290

Brain Builders

1. 44,600
2. 406,060
3. 66,963
4. 78,720
5. 21,699
6. 33,000
7. 45,375
8. 569,688

Trick 50

Elementary

1. 3,885
2. 5,994
3. 2,553
4. 8,991
5. 4,773
6. 7,992
7. 6,882
8. 1,998
9. 2,886
10. 5,883
11. 4,995
12. 6,771
13. 1,665
14. 3,663

Brain Builders

1. 27,775
2. 39,996
3. 46,662
4. 57,772
5. 69,993
6. 14,443
7. 23,331
8. 35,552

Trick 51

Elementary

1. 34,632
2. 15,873
3. 47,841
4. 37,962
5. 24,642
6. 59,496
7. 78,144
8. 40,515
9. 46,398
10. 92,019
11. 73,593
12. 24,309
13. 59,496
14. 86,247

Brain Builders

1. 136,974
2. 458,652
3. 355,977
4. 462,426
5. 828,060
6. 213,786
7. 355,089
8. 664,668

Trick 52

Elementary

1. 949,050
2. 824,175
3. 729,270
4. 479,520
5. 925,074
6. 298,701
7. 152,847
8. 607,392
9. 89,910
10. 74,925
11. 65,934
12. 84,915
13. 70,929
14. 88,911

Brain Builders

1. 6,993
2. 12,987
3. 26,973
4. 7,992
5. 35,964
6. 16,983
7. 40,959
8. 5,994

Trick 53

Elementary

1. Definitely incorrect
2. Probably correct
3. Probably correct
4. Definitely incorrect
5. Definitely incorrect
6. Probably correct
7. Definitely incorrect
8. Probably correct
9. Definitely incorrect
10. Probably correct
11. Definitely incorrect
12. Probably correct
13. Probably correct
14. Definitely incorrect

Brain Builders

1. Probably correct
2. Definitely incorrect
3. Definitely incorrect
4. Probably correct
5. Definitely incorrect
6. Probably correct
7. Probably correct
8. Definitely incorrect

Trick 54

Elementary

1. Probably correct
2. Probably correct
3. Definitely incorrect
4. Probably correct
5. Definitely incorrect
6. Definitely incorrect
7. Probably correct
8. Definitely incorrect
9. Probably correct
10. Probably correct
11. Definitely incorrect
12. Definitely incorrect
13. Probably correct
14. Probably correct

Brain Builders

1. Definitely incorrect
2. Probably correct
3. Probably correct
4. Definitely incorrect
5. Probably correct
6. Definitely incorrect
7. Probably correct
8. Definitely incorrect

Quick Quizzes

Week 1

Elementary

1. 584 (Trick 1)
2. 2,491 (Trick 2)
3. 407 (Trick 3)
4. 14 (Trick 4)
5. 57 (Trick 5)
6. 182 (Trick 6)
7. 18,312 (Trick 7)
8. 600 (Trick 8)
9. 102 (Trick 9)
10. 24 (Trick 10)
11. 277 (Trick 11)
12. $19.73 (Trick 12)
13. 2,170 (Trick 13)

Brain Builders

1. 3,766 (Trick 1)
2. 3,216 (Trick 3)
3. 15 (Trick 4)
4. 78,660 (Trick 7)
5. 66 (Trick 8)
6. 330 (Trick 9)
7. $36.57 (Trick 12)

Week 2

Elementary

1. 516 (Trick 14)
2. 39 (Trick 15)
3. 280 (Trick 16)
4. 948 (Trick 17)
5. 42 (Trick 18)
6. 54 (Trick 19)
7. 615 (Trick 20)
8. 312 (Trick 21)
9. 156 (Trick 22)
10. 18 (Trick 23)
11. 98 (Trick 24)
12. 57 (Trick 25)
13. 178 (Trick 26)

Brain Builders

1. 283 (Trick 14)
2. 518 (Trick 16)
3. 32 (Trick 18)
4. 2,835 (Trick 20)
5. 540 (Trick 22)
6. 19 (Trick 24)
7. 424 (Trick 26)

Week 3

Elementary

1. 221 (Trick 27)
2. 441 (Trick 28)
3. 784 (Trick 29)
4. 1,089 (Trick 30)
5. 231 (Trick 31)
6. 216 (Trick 32)
7. 676 (Trick 33)
8. 1,156 (Trick 34)
9. 15,710 (Trick 35)
10. 1,188 (Trick 36)
11. 9,409 (Trick 37)
12. 11,236 (Trick 38)
13. 9,118 (Trick 39)

Brain Builders

1. 304 (Trick 27)
2. 5,184 (Trick 29)
3. 990 (Trick 31)
4. 5,776 (Trick 33)
5. 42,918 (Trick 35)
6. 990,025 (Trick 37)
7. 991,014 (Trick 39)

Week 4

Elementary

1. 875 (Trick 40)
2. 525 (Trick 41)
3. 103,020 (Trick 42)
4. 54 (Trick 43)
5. 1,345 (Trick 44)
6. 211 (Trick 45)
7. 1,804 (Trick 46)
8. 2,516 (Trick 47)
9. 48 (Trick 48)
10. 3,560 (Trick 49)
11. 4,995 (Trick 50)
12. 25,641 (Trick 51)
13. 749,250 (Trick 52)

Brain Builders

1. 9,975 (Trick 40)
2. 285,418 (Trick 42)
3. 78 (Trick 43)
4. 1,672 (Trick 46)
5. 200 (Trick 48)
6. 68,882 (Trick 50)
7. 32,967 (Trick 52)

Solutions to Final Exam

1. 273 (Trick 1)
2. 2,190 (Trick 2)
3. 235 (Trick 3)
4. 11 (Trick 4)
5. 39 (Trick 5)
6. 96 (Trick 6)
7. 18,522 (Trick 7)
8. 2,100 (Trick 8)
9. 144 (Trick 9)
10. 19 (Trick 10)
11. 240 (Trick 11)
12. $23.72 (Trick 12)
13. 1,470 (Trick 13)
14. 357 (Trick 14)
15. 77 (Trick 15)
16. 441 (Trick 16)
17. 1,308 (Trick 17)
18. 29 (Trick 18)
19. 42 (Trick 19)
20. 496 (Trick 20)
21. 343 (Trick 21)
22. 168 (Trick 22)
23. 25 (Trick 23)
24. 99 (Trick 24)
25. 59 (Trick 25)
26. 165 (Trick 26)
27. 252 (Trick 27)
28. 961 (Trick 28)
29. 324 (Trick 29)
30. 529 (Trick 30)
31. 396 (Trick 31)
32. 360 (Trick 32)
33. 1,296 (Trick 33)
34. 576 (Trick 34)
35. 31,884 (Trick 35)
36. 490 (Trick 36)
37. 9,604 (Trick 37)
38. 11,881 (Trick 38)
39. 9,215 (Trick 39)
40. 1,575 (Trick 40)
41. 1,125 (Trick 41)
42. 143,820 (Trick 42)
43. 300 (Trick 43)
44. 1,763 (Trick 44)
45. 224 (Trick 45)
46. 2,112 (Trick 46)
47. 2,736 (Trick 47)
48. 60 (Trick 48)
49. 2,064 (Trick 49)
50. 7,992 (Trick 50)
51. 46,953 (Trick 51)
52. 639,360 (Trick 52)
53. Definitely incorrect (Trick 53)
54. Probably correct (Trick 54)

Parlor Tricks

Parlor Trick 3

A. 6,809,332 B. 31,950,666 C. 59,631,138 D. 13,232,076

Parlor Trick 4: Basic Version

A. 6	C. 80	E. 29	G. 95	I. 42	K. 7
B. 61	D. 54	F. 13	H. 78	J. 38	L. 99

Parlor Trick 4: Expanded Version

A. 100	C. 143	E. 214	G. 105	I. 158	K. 117
B. 137	D. 176	F. 199	H. 222	J. 240	L. 251

Number Potpourris

15. $2^{88} - 1$ seconds, or about 9.8 quintillion years (that's 700 million times the estimated age of the universe!).

17. If you were determining the number of different ways, say, five books could be arranged on a shelf, you would take $1 \times 2 \times 3 \times 4 \times 5$ (described as "5 factorial"), which equals 120 different ways. So to figure out the number of ways you could arrange 54 tricks, you would take $1 \times 2 \times 3 \times \ldots \times 54$. This product turns out to be a number so large, it contains 72 digits!

Number Challenges

1. 1. b (136°F)
   2. a (−128°F) (*Note:* −459°F is absolute zero!)

2. 1. c (110 stories)
   2. d (6.75 miles)

3. 1. c (4.5 billion years)
   2. b (14 billion years)

4. 10 yards square (not 5 yards square!)

5. 9^{9^9} (9 raised to the 9th power of 9). It is a number so large, it contains nearly 370 million digits!

Summary of the 54 Number-Mastery Tricks for Handy Reference

TRICK 1: When multiplying by a one-digit number, enter each product without carrying, moving one column to the left each time.

TRICK 2: When adding a column of numbers, "bunch up" the partial sums without carrying, moving one column to the left each time.

TRICK 3: Where possible, subtract without borrowing by handling two columns at once.

TRICK 4: When dividing with even numbers, cut each number in half to simplify the computation.

TRICK 5: When adding pluses and minuses, you may (a) proceed one number at a time, (b) add the pluses, then the minuses, and combine, or (c) add, netting out the minuses against the pluses.

TRICK 6: When adding numbers that are close to each other, guess at the midpoint, multiply, then compute with the distances from the midpoint.

TRICK 7: Where possible, convert a two-digit multiplier into two one-digit multipliers. For example, 384×42 could be computed as $384 \times 7 \times 6$.

TRICK 8: When multiplying three or more numbers, proceed in a different order if it will simplify the computation.

TRICK 9: Multiply by 6 by multiplying by 3, then 2 (or by 2, then 3). For example, 18×6 could be computed as $18 \times 3 \times 2$.

TRICK 10: Divide by 6 by dividing by 3, then 2 (or by 2, then 3). For example, $102 \div 6$ could be computed as $102 \div 2 \div 3$.

TRICK 11: When adding the same digit several times in a column, multiply to speed up the calculation.

TRICK 12: When adding numbers just under multiples of $1, round up, add, then deduct the amounts that were "added on."

TRICK 13: When multiplying, look for "digit-multiples" to speed up the computation.

TRICK 14: When dividing, multiply and subtract in your head to conserve both space and time.

TRICK 15: Where possible, convert a two-digit divisor into two one-digit divisors. For example, 4,312 ÷ 56 could be computed as 4,312 ÷ 8 ÷ 7.

TRICK 16: Multiply a two-digit number by a one-digit number by working from left to right, focusing on the "place value" of each digit.

TRICK 17: Multiply a three-digit number by a one-digit number by working from left to right, focusing on the "place value" of each digit.

TRICK 18: Where possible, treat a number being divided as 100 (or a multiple of 100) plus something. For example, 115 ÷ 5 could be computed as (100 + 15) ÷ 5, or (100 ÷ 5) + (15 ÷ 5).

TRICK 19: Where possible, split a number being divided into two parts, each of which can be divided evenly by the divisor. For example, 105 ÷ 3 could be computed as (99 + 6) ÷ 3, or (99 ÷ 3) + (6 ÷ 3).

TRICK 20: To multiply a number by 21, 31, or the like, multiply by 1 less (than the 21, 31, etc.), then add the number. For example, 45 × 21 could be computed as (45 × 20) + 45.

TRICK 21: To multiply a number by 19, 29, or the like , multiply by 1 more (than the 19, 29, etc.), then subtract the number. For example, 7 × 29 could be computed as (7 × 30) − 7.

TRICK 22: Multiply by 12 by breaking the 12 into smaller parts. For example, 16 × 12 could be computed as 16 × 6 × 2.

TRICK 23: Divide by 12 by breaking the 12 into smaller parts. For example, 156 ÷ 12 could be computed as 156 ÷ 2 ÷ 2 ÷ 3.

TRICK 24: Where possible, treat a number being divided as 100 (or a multiple of 100) minus something. For example, 92 ÷ 4 could be computed as (100 − 8) ÷ 4, or (100 ÷ 4) − (8 ÷ 4).

TRICK 25: Where possible, treat a number being divided as something minus something, where both numbers can be divided evenly by the divisor. For example, 144 ÷ 3 could be computed as (150 − 6) ÷ 3, or (150 ÷ 3) − (6 ÷ 3).

TRICK 26: When subtracting numbers just under 100 (or multiple of 100), subtract the next higher multiple of 100, then add back the amount oversubtracted.

TRICK 27: To multiply with the numbers 11 through 19, compute as in the following example: 17 × 12 = [(17 + 2) × 10)] + (7 × 2).

TRICK 28: To square a number ending in 1, compute as in the following example: 31^2 = (32 × 30) + 1. To square a number ending in 9, compute as in the following example: 19^2 = (18 × 20) + 1.

TRICK 29: To square a number ending in 2, compute as in the following example: $32^2 = (34 \times 30) + 4$. To square a number ending in 8, compute as in the following example: $28^2 = (26 \times 30) + 4$.

TRICK 30: To square a number ending in 3, compute as in the following example: $23^2 = (26 \times 20) + 9$. To square a number ending in 7, compute as in the following example: $17^2 = (14 \times 20) + 9$.

TRICK 31: To multiply by a multiple of 11, multiply by the previous multiple of 10, then add 10%.

TRICK 32: To multiply by a multiple of 9, multiply by the next multiple of 10, then subtract 10%.

TRICK 33: To square a number ending in 6, compute as in the following example: $16^2 = 15^2 + 15 + 16$.

TRICK 34: To square a number ending in 4, compute as in the following example: $24^2 = 25^2 - 25 - 24$.

TRICK 35: Multiply a four-digit number by a one-digit number by working from left to right, focusing on the "place value" of each digit.

TRICK 36: Multiply a two-digit number by a two-digit number by multiplying tens digits, cross-multiplying twice, then multiplying ones digits. Then add the products.

TRICK 37: To square a number between 90 and 100, subtract from the number its distance from 100 for the left half of the answer. Then square the distance for the right half.

TRICK 38: To square a number between 100 and 110, add to the number its distance from 100 for the left portion of the answer. Then square the distance for the right portion.

TRICK 39: To multiply two numbers that are a little under 100, subtract from either number the distance from 100 of the other number for the left half of the answer. Then multiply the two distances together for the right half.

TRICK 40: To multiply two consecutive numbers ending in 5, square the number exactly in the middle and subtract 25. For example, $15 \times 25 = 20^2 - 25$.

TRICK 41: To multiply two alternate numbers ending in 5, square the number exactly in the middle and subtract 100. For example, $35 \times 55 = 45^2 - 100$.

TRICK 42: To multiply two three-digit numbers whose middle digits are zero (e.g., 409×703), multiply hundreds digits, cross-multiply twice, add the products, then multiply the ones digits. Then combine the amounts from Steps 1, 3, and 4.

TRICK 43: To multiply with one number just under a whole, round up the mixed number, multiply, then subtract the overstated amount.

TRICK 44: When adding a column of numbers, place at the bottom of the next column each carried amount. A new digit of the answer will appear as each column is added.

TRICK 45: When adding two numbers, break apart one number so that one of its components plus the other number will total 100 (or a multiple of 100).

TRICK 46: To multiply two two-digit numbers consisting of a number with a repeated digit (as 66) and another number whose digits add to 10 (as 37), first pretend that the latter number is 10 greater than it is (47, using our example above). Then multiply the tens digits together, then the ones digits together, writing the answer from left to right.

TRICK 47: To multiply two two-digit numbers whose tens digits add to 10 and whose ones digits are identical (e.g., 76 × 36), multiply the tens digits together and add the ones digit for the left half of the answer. Then square the ones digit for the right half.

TRICK 48: When multiplying a whole number by a fraction (e.g., $28 \times \frac{3}{4}$), move the denominator of the fraction underneath the whole number if it will simplify the calculation.

TRICK 49: Where manageable, multiply by two or three digits at once. For example, when computing 812 × 5, multiply 12 by 5 in one fell swoop.

TRICK 50: To multiply a two-digit number by 111, add the digits of the number, and insert the sum within the number itself twice. This trick will only work for digit-sums of 9 or less.

TRICK 51: To multiply a three-digit or larger number by 111, first carry down the ones digit of the number. Then add the ones and tens digits; the ones, tens, and hundreds digits; and so forth. Finally, carry down the first digit of the number. Carry when necessary.

TRICK 52: To multiply a one-, two-, or three-digit number by 999, subtract 1 from the number, and affix the difference between 1,000 and the number.

TRICK 53: To check addition and subtraction, obtain a "digit total" for each number by alternating between subtraction and addition of digits as you work from right to left. Then compare the digit-total sum (for addition) or digit-total difference (for subtraction) with the digit total of the answer you are checking.

TRICK 54: To check multiplication, obtain digit totals as explained in Trick 53. Then compare the digit-total product with the digit total of the answer you are checking. To check division, treat as multiplication, and check in the same manner.